OXFORD BIOLOGY PRIMERS

Discover more in the series at
www.oxfordtextbooks.co.uk/obp

Published in partnership with the Royal Society of Biology

POWER ANALYSIS:
AN INTRODUCTION FOR THE LIFE SCIENCES

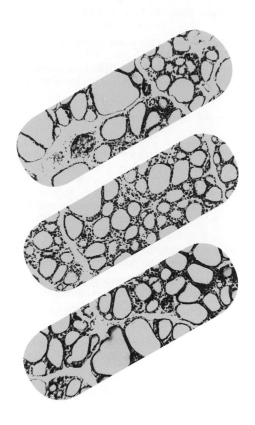

POWER ANALYSIS: AN INTRODUCTION FOR THE LIFE SCIENCES

Nick Colegrave and Graeme D. Ruxton

OXFORD
UNIVERSITY PRESS

Great Clarendon Street, Oxford, OX2 6DP,
United Kingdom

Oxford University Press is a department of the University of Oxford.
It furthers the University's objective of excellence in research, scholarship,
and education by publishing worldwide. Oxford is a registered trade mark of
Oxford University Press in the UK and in certain other countries

Impression: 1

Published in the United States of America by Oxford University Press
198 Madison Avenue, New York, NY 10016, United States of America

British Library Cataloguing in Publication Data
Data available

Library of Congress Control Number: 2020948545

ISBN 978-0-19-884663-5

Printed in Great Britain by
Bell & Bain Ltd., Glasgow

PREFACE

Welcome to the Oxford Biology Primers

There has never been a more exciting time to be a biologist. Not only do we understand more about the biological world than ever before, but we're using that understanding in ever-more creative and valuable ways.

Our understanding of the way our genes work is being used to explore new ways to treat disease; our understanding of ecosystems is being used to explore more effective ways to protect the diversity of life on Earth; our understanding of plant science is being used to explore more sustainable ways to feed a growing human population.

The repeated use of the word 'explore' here is no accident. The study of biology is, at heart, an exploration. We have written the Oxford Biology Primers to encourage you to explore biology for yourself—to find out more about what scientists at the cutting edge of the subject are researching, and the biological problems they're trying to solve.

More than anything, however, we hope this series will reveal to you, its readers, that biology is awe-inspiring, both in its variety and its intricacy, and will drive you forward to explore the subject further for yourself.

ABOUT THE AUTHORS

Nick Colegrave, Professor of Experimental Evolution, University of Edinburgh
Graeme D. Ruxton, Professor of Biology, University of St Andrews

ACKNOWLEDGEMENTS

Lucy Wells and Sophie Ladden at OUP were great guides in the initial development of the book here. They found six fantastic reviewers, whose perceptive comments really stimulated significant improvements in the book. Of these six, we are particularly grateful to Innes Cuthill, who provided an exceptionally in-depth critique of our first draft material, and was the perfect balance between supportive and challenging.

Both authors are rather graphically challenged, and the figures in this book benefited hugely from the imagination and technical skills of Rosalind Humphreys.

The text was greatly improved by comments from Anna Rouvi`ere.

Kat Keogan read through all chapters and checked all scripts, spotting many mistakes and making valuable suggestions on the way.

Andrew Beckerman and Owen Petchey taught us how to reproduce R code and output in a much more effective document form that we would have managed on our own.

CONTENTS

INTRODUCTION: WHY SHOULD YOU READ THIS BOOK?

Part I: Why you should want to do power analysis

If we wanted to persuade you to read this book, we could try and scare you. We could say that your career as a scientist is really going to be held back if you don't embrace power analysis. Funding is getting harder and harder to secure, funders are increasingly asking you specifically to justify the power of your experiments, and you really can't afford to handicap your grant applications by not doing a professional job of justifying your experimental designs in terms of power. For biologists in particular, gaining ethical approval for experiments is often an essential precursor to any data collection; and ethics boards increasingly expect you to demonstrate that you have thought of the power of your experiments. Often gaining ethical approval will require you to demonstrate that your research uses the minimum number of humans, animals, or biological material to give a clear answer to your questions. Discussion of statistical power will be integral to any such demonstration. Finally, the limitations of underpowered experiments are increasingly appreciated amongst scientists, including potential editors and referees of manuscripts that you might be sending to scientific journals. Convincing others that your experiment is not underpowered is going to increase the chances of your work being published where you would most like it, and indeed increase the chances of it getting published at all. Scientific journals also pride themselves on not publishing unethical work, so they would take a dim view of an experiment that uses more biological material than it needs to in order to have a good chance of giving definitive answers to the scientific questions asked. All of the foregoing is true, but we are just too nice as people to try and scare you into reading our book. We would rather open your eyes to how rewarding power analyses can be.

Your career in science is likely going to involve designing data collection exercises (which we'll call experiments for simplicity), so you can use those data to address interesting questions that you want to answer. Collecting data can often be a hard slog, painstakingly and repetitively doing the same thing again and again. Your motivation to knuckle down to data collection is going to be higher if you are really confident that: (i) you are collecting the data in the most efficient way you can; (ii) you know how you are going to analyse your data once you have it; and (iii) that analysis has a really good chance of offering definitive answers to the questions that matter to you. Carrying out a good power analysis before you start your data collection can deliver all three of these for you. Further, there is no experiment that you can imagine for which the methods we discuss here can't offer you a power analysis; and that analysis is not going to involve you in a lot of extra work. Better yet, the little extra work required should be intellectually engaging and sometimes even fun (if you are wired like we are).

Yet another way to motivate power analysis is that prevention is better than cure. Power analysis requires a bit of extra work at the start of your investigation—but it should clarify and simplify the collection and analysis of your data, and your writing up of those data. If you skip the power analysis then you are much more likely to end up with an inconclusive, unsatisfying data set that will require you to get much more creative in your statistical analysis and 'spinning' of your results to try and get something worthwhile out of all your efforts. Investment in power analysis at the start (when you are energized, motivated, and not time-stressed) is going to spare you a lot of pain, frustration, and disappointment at the end of your study (when energy, motivation, and time may not be in such ample supply).

Part 2:   Why you should want to do power analyses our way

We want to encourage you to perform power analyses by simulation using computer code that you write yourself. But there are alternatives: there are purpose-designed power analysis software packages you could use instead. We favour our approach for several reasons, but let's cover the most important reason first—because we think it's a clincher. Simply put, if you absorb the conceptual and practical advice that we cover in this slim volume then you will be equipped to carry out power analysis on any experiment at all that you can imagine. That is not going to be true for any off-the-shelf power-analysis package. Those packages can only be set up to allow you to explore a pre-scribed list of different designs—that list will be longer and shorter for different packages, but it will always be finite. These limitations will exist not only in terms of the design decisions the package allows you to make, but also the range of biological and practical situations it can cover. With our approach you will be bounded only by your own creativity in designing experiments—not by what was easy for some software developer. Stick with us and you can explore the power of any experiment you can imagine, to answer any question you can imagine about any underlying biological system you might encounter. No package can come even close to offering that. If that was the only benefit to doing it our way, that should be enough. But we are not finished yet.

Doing power analyses our way involves you really understanding your experiment fully. You know that with off-the-shelf packages you can plug in numbers, and an answer comes out, and you might not necessarily know how that number was calculated and what it means. There is no danger of that with our approach. You describe the experiment you want to explore, the data you will collect from the experiment, and the analyses you will perform on the experiment fully, so all the assumptions and design decisions are there for you and anyone else to see. We think this will help you understand your experiments better, and so design better experiments and explain and justify your design decisions more clearly and conclusively to others.

With any new off-the-shelf package there is a lead time required to learn how to use it. If it is a package you don't use that often, then you may need to keep re-learning it again and again. Whereas our approach involves some really pretty simple extensions to the programming language that you use to analyse and present your data (most obviously R to an increasing fraction of scientists). Hence, our method should not be substantially harder to implement than the alternatives, and might often actually be easier.

Lastly, our method involves you specifying how you will analyse your data as part of your power analysis. This delivers a massive saving at the end of your project. The R code that you use in your power analysis will need only trivial changes to provide a stress-free analysis of your actual data once you have collected them. You can look at this benefit as a means of mitigating the investment in power analysis. You can also see this as a way of moving decisions about data analyses from the end to the start of the investigation—when you have more energy and time. You can also see it as a way to strengthen grant applications and ethical approval submissions through demonstrating that you have your analyses worked out early.

Why have we based the book on R and how much R will we assume you know?

Pretty simply, R is growing and growing as a popular package for performing data analysis, and we think that trend will only continue in the coming years. In addition, the advent of the RStudio interface has made interacting with R substantially more straightforward and less intimidating. We don't assume you are especially proficient in R or RStudio, but we do imagine that this book will go more smoothly for you if you already have some familiarity with using R. If you haven't used R before then we don't think this book will be a dead loss to you. We have tried to explain enough of the bare bones as we go along (and we have also provided a quick start guide to get you proficient in the basics of RStudio, which can be found in the electronic supplementary material associated with this introduction). However, this is not the best book to introduce you to statistical analysis in R, nor is it a book about statistical methodology or experimental design. We do assume you already have been doing a little designing of studies and analysing data from them before you decide that power analysis is something you want to be proficient in. But having made that decision, you have identified the ideal book to help you ☺

How to use this book

Any way you like. We think there is real value to be had from reading it cover-to-cover like a novel, especially if power analysis is relatively new to you. But we appreciate that you are busy—that's why we have kept the book as compact as we possibly can. If you are a little bit more experienced in thinking about statistical power, or have a very specific issue you want to resolve in your head then by all means just dip in. We have slaved over the index to make this as easy as we can for you, as well as making the headings as informative as we can, defining key terms when we first use them, and cross-referencing so you can knit ideas together without reading cover-to-cover. To help you get orientated—here is a brief outline of the contents of the book:

Structure of the book

The book is split into chapters, and these chapters are chopped up into subsections. We feel there is a logical ordering within and between chapters. However, there was some material that we felt would be useful to some but not all readers, and we have abstracted that from the main structure of the book. Material which is confirmatory—enriching understanding of concepts that we cover briefly in the main text—we have labelled 'case studies'. You could safely skip this material if you feel the main text meets your needs. But where you would like to test and strengthen your understanding then the case studies should help. If the material is related to concepts we discuss, and we feel can enrich the understanding of statistical power for some readers, but is not really primarily concerned with statistical power, then we abstract that material out of the book altogether and place it in the Electronic Supplementary Material (ESM for short) that you can access on the website linked to this book. You can also find electronic versions of all of the R Scripts from the book in the ESM associated with the appropriate chapter.

Chapter 1: **What is statistical power?**

Statistical power has a specific technical definition, and the reader should leave this chapter with a strong understanding of what exactly power is.

Chapter 2: **Why low power is undesirable**

Here we explain why statistical power is a useful concept, and the multiple reasons why low power is undesirable in a scientific study.

Chapter 3: **Improving the power of an experiment**

Here we explain the multiple interacting factors that influence the power of an experiment, and thus the issues a researcher can consider when attempting to increase the power of their study.

Chapter 4: **How to quantify power by simulation**

Here we explain conceptually how to calculate the power of any experiment, and how practically this can be done for simple one-factor experiments with some easy R code.

Chapter 5: **Simple factorial designs**

Here we show how conceptual and practical understanding from Chapter 4 can be generalized easily to evaluate and compare the powers of more complex multi-factorial experiments.

Chapter 6: **Extensions to other designs**

Here we show how our understanding from Chapters 4 and 5 can be generalized to allow you to calculate the power of any study design you can imagine.

Chapter 7: **Dealing with multiple hypotheses**

Often you will want to address a number of related questions in a single study—here we talk about some things to think about to ensure that you have good power to address all of them.

Chapter 8: **Applying our simulation approach beyond null hypothesis testing: parameter estimation, Bayesian, and model-selection contexts**

To this point we have discussed statistical power in the context of the inferential statistical approach of testing null hypotheses; here we demonstrate how naturally everything we have discussed translates into parameter estimation, Bayesian, and model-selection approaches.

Appendix: **Some handy hints on simulating data in R**

A practical resource for readers starting to calculate power for their own studies.

1 WHAT IS STATISTICAL POWER?

Learning objectives

By the end of the chapter, you should be able to:

- Define the terms 'population' and 'sample' in the context of a scientific study.
- Identify the null hypothesis associated with any scientific question.
- Describe precisely what inferences can and cannot be drawn from a p-value.
- Explain the concepts of Type I and Type II error in terms of null hypothesis testing.
- Define statistical power and explain its link to Type II error.

Since this whole book is about **statistical power**, things will go terribly wrong if we do not make crystal clear what we mean by statistical power. So our sole ambition for this chapter is to clearly define statistical power. To achieve this, we realized that we first of all have to explain some fundamental concepts about statistical testing. Once you have a good grasp of how statistical testing works, it turns out that defining statistical power is very easy. So this chapter is a careful build up to a really short and simple definition of statistical power given in the last section of the chapter. If you already feel you are pretty familiar with statistical testing, and even statistical power, then jump to our final section and see if our definition is completely clear to you. If it is, then it's probably safe to move on. If not, you will want to delve back into this chapter a little.

1.1   An important preliminary: sampling and statistical testing

The five most common causes of death for men and women in the United Kingdom are listed in Table 1.1. Imagine that you want to know the typical causes of death of males in Scotland (a subject of strong interest to the authors). To get a definitive answer to this question you could somehow categorize the causes of death noted on official death certificates of every single male who died in Scotland over the last year. This would obviously be a huge study, and in reality would probably not be practical without very substantial effort (given that there are over 25,000 males a year dying in Scotland). If you do want a good answer to this question, but are not prepared to devote a whole year of your life to collecting the data, then another approach is required. The approach we are going to outline makes use of two important statistical concepts that are central to the approaches in this book, populations and samples, so let's be clear what we mean by them right from the start.

When statisticians talk about a population they mean the collection of all the entities that could be evaluated to answer the question of interest. So, for your question about cause of death in Scottish males, your statistical population would be all the males who die in Scotland within some specified span of time. Measuring every individual in the population should give you a definitive answer to your research question. However, as we have already pointed out, unless you happen to work on a highly endangered species, it is not usually practical to measure all the individuals in the entire population. Instead, scientists study an appropriate representative sample of the population of interest and then make use of statistics to try to make generalizations about the population that they represent. That is, rather than exhaustively measuring every individual in the population, you measure a smaller subset of those individuals. If that subset is a representative sample of the whole population then we would hope that features that we see in the data collected from the subset should be broadly similar to (but not necessarily absolutely identical to) those we would have found if we had measured the whole population.

Table 1.1 The top five causes of death for men and women in the UK and the percentages of deaths where each of these causes was considered the main factor.

UK male deaths	%	UK female deaths	%
Heart disease	14.2	Dementia and Alzheimer's disease	15.3
Dementia and Alzheimer's disease	8.0	Heart disease	8.8
Lung cancer	6.5	Stroke	7.5
Chronic lower respiratory diseases	6.2	Influenza and pneumonia	6.0
Stroke	5.6	Chronic lower respiratory diseases	6.0

Data from 2018 UK Government, Office of National Statistics.

Statistical theory has evolved to offer us tools for estimating how similar we expect features seen in the sample to be to equivalent features in the population. As an example, if we measured a representative sample of 250 deaths of Scottish males and found that 8 per cent of them occurred as a result of an accident then we would turn to statistics to guide us as to what that tells us about the whole population of deaths of Scottish males that we are interested in. Would we expect that exactly 8 per cent of the whole population involved accidental deaths, or might that fraction actually be a little more or less? Is it likely to be as low as 7 per cent? Or 5 per cent?

So we might measure the cause of death of an appropriate sample of 250 males dying in Scotland, and use that information to try to say something general about Scottish males. The important thing is that we are not specifically interested in the 250 particular subjects that we happened to study, but in what they can tell us about the population that they represent. There is nothing special about the 250 individuals that we happened to choose for our sample; we might equally as well have chosen an entirely different 250 individuals.

Now how much our sample can tell us about the population will depend on how representative that sample is, and this leads to an important trade-off—the most representative sample is the entire population. Measuring the entire population gives us a definitive answer but is impractical; measuring a much smaller sample is more practical but does not give us a definitive answer to the question we were originally interested in—rather it gives us evidence that is not definitive but suggestive. Generally, the larger the sample we measure, the more representative it will be, and the more confident we can be about our conclusions. Statisticians have spent a great deal of time over the last 200 years developing tools that allow us to use information from our samples to draw inferences about the populations that they represent. These tools allow us to quantify how strongly suggestive the data from our sample is about what is actually occurring in the bigger population, which is actually what we are interested in.

1.1.1 Defining populations in an experimental setting

Whilst at first sight it may not seem as obvious, we can think of experiments in terms of populations and samples too. The use of multi-vitamin supplements is less common in children than adults (Fig. 1.1). Suppose you are carrying out a study to examine the effect of a multi-vitamin supplement on children's health. Your planned study has a control group of individuals who take a placebo pill daily, and an experimental group who take the multi-vitamin; and you are going to monitor the number of days they are absent from school over the length of study as a proxy for ill-health. The samples are the groups of individual subjects that you assign to each treatment group. So in your study you have two samples; the first is made up of the individuals taking the multi-vitamin supplement, and the second made up of the individuals in the control group. But what are your populations? Here things get a little more conceptual, because unlike the real population of death of males in Scotland, there is not really an equivalent real population called 'all children taking the multi-vitamin supplement'. Instead, you have to imagine a hypothetical population of all the individuals that could have been included in your trial and could then have been randomized into the treatment group. You can then use your sample of vitamin-taking children to try to say something about the

Fig. 1.1 Percentages of US children and adults regularly taking multi-vitamin supplements.

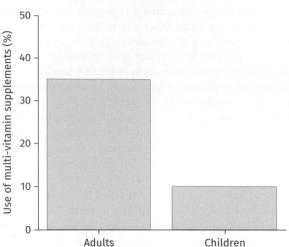

Source: Rock, C.L., 2007. Multi-vitamin-multimineral supplements: who uses them? *The American Journal of Clinical Nutrition*, 85(1), pp. 277S–279S.

hypothetical population it represents. For example, what is the average number of days missed from school each year through ill-health of individuals in that hypothetical population, and how does this compare to the same measure of other hypothetical populations (for example, the population of all individuals that could be given the placebo in our study)? If the notion of hypothetical populations feels a little too abstract for you, don't worry too much. The important point is that we can think of any experiment as an exercise in drawing samples from populations. This is a concept we will use extensively in later chapters as a way to estimate the power of our studies.

 Key points

- We are interested in understanding large populations, but we end up doing experiments on small samples from those populations.
- The behaviour, or traits, of these samples should be a reasonably reliable guide to the answers to population-level questions.
- Statistical testing offers a way of quantifying how reliable such small-sample experiments are as guides to population-level effects.

1.2   Null hypothesis statistical testing

For much of this book we will couch power analysis in terms of what is called null hypothesis statistical testing. There are other approaches to inferential statistics (see section 1.2.1) that power analysis is just as relevant to. We will defer exploration of them until Chapter 8 and focus on null hypothesis statistical testing, because it remains the most commonly-encountered form of inferential statistics, even if the others are steadily gaining ground. And the skills you will learn in dealing with null hypothesis statistical testing can easily be applied more generally to other approaches.

1.2.1 Different forms of inferential statistics

Remember that any experiment involves measurements of a sample drawn from the larger population that we are actually interested in. Descriptive statistics are a means of summarizing the data collected on the sample. Imagine if we still want to test whether multi-vitamin supplements provide health benefits to school children. We might recruit 200 children into our study that have not previously taken a supplement and randomize them into one of two groups, wherein they take either the supplement or a placebo once a day for the next school year. For each individual we could measure poor health as the number of days that they were recorded as missing school through ill-health. An example descriptive statistic in this study might be the difference in the two average numbers of schooldays missed by children in our two 100-pupil sample groups.

Inferential statistics are approaches that help us make inferences about the population from our observation of the sample. There are three fundamental forms of such inference: null hypothesis testing, parameter estimation and model selection. We explain null hypothesis testing in the main text but this often essentially boils down to describing how surprising any difference we find between treatment groups in our sample would be, if in fact there was no difference at the population level. For example, imagine we found that children in our placebo sample have 10 days per year off school on average through ill-health, and for those provided with the multi-vitamin this number was 8 days on average. Null hypothesis tests ask how plausible these data seem if we make the assumption that actually the multi-vitamin has no effect. Key to evaluating this 'plausibility' is calculation of a p-value. The p-value is the probability of observing your data or even more extreme data if the null hypothesis were actually true. In our case our samples suggest a two-day difference on average between the two groups of children. In this case the p-value would be the probability of seeing a two-day difference (or an even larger difference) just through the noise (see section 3.1) introduced by sampling when actually the multi-vitamin had exactly the same effect as the placebo.

In the parameter estimation approach, a parameter is simply any attribute of the underlying population that we are interested in. In our case, the parameter we are interested in is the reduction in number of days off school associated with taking a daily multi-vitamin. Parameter estimation uses information from the sample to estimate what range of values for that parameter, at a population level, might seem plausible in the light of its measured value in a small sample. In our example, the difference in sample averages was exactly 2 days. Parameter estimation techniques could provide guidance on, for example, whether these data suggest that at the level of whole populations this difference might plausibly be as much as, say, 2.1 days, or 2.8 days.

Model selection techniques compare a battery of two or more models that aim to describe a child's attendance at school over the year. In our example, there are two possible explanations for the data we obtained: one in which there is no effect of multi-vitamins, and the observed difference between our two sample groups is due to chance, and one in which the averages of the two sample groups differ because multi-vitamins actually produce a decrease in days off school. Statisticians would call these two candidate explanations 'models', and are often interested in comparing how well different models account for the data obtained. In our case, our models differ in how strongly they weigh taking a multi-vitamin supplement relative to other factors. Inference on how important the multi-vitamin supplement seems to be can be made in terms of how well relative to each other these different models seem to be able to explain the data we observed in our experiment. For example, we might fit a statistical model that tries to estimate the number of days off school each of the children has on the basis of their age and sex only. We might then fit a model that considers age, sex and which of our two treatment groups (multi-vitamin or placebo) that child belongs to. We would then ask whether the second model fits our data better, and by how much.

None of these three inferential methodologies is inherently superior to any other. They offer complementary ways of quantifying the uncertainty that inevitably arises when we try and make inferences on a whole population from examining just a sample. The relative merits of these different approaches will vary according to the nature of the experiment and the nature of the questions you want to ask about the population—but no matter their relative merits in any one situation, they all have some use. Thus, researchers tend to specialize in using one particular approach, because using one approach skilfully in any particular situation is normally more valuable than agonizing over which particular technique is inherently the most effective.

1.2.2 What exactly is a null hypothesis?

In order to understand null hypothesis statistical testing we need to understand what a null hypothesis is. For any biological question that we are interested in answering, the null hypothesis is usually the answer that nothing is happening. If we ask 'does taking a multi-vitamin supplement alter a child's rate of school attendance?' then the null hypothesis is that 'taking a multi-vitamin supplement does not affect a child's rate of school attendance'. If we ask, 'across the different species of mammal, what is the relationship between characteristic body size and characteristic longevity?' then the null hypothesis is 'across the different mammal species, body size and longevity are totally independent and unrelated.' More examples follow.

> **Question:** Do the sexes of salmon differ in parasite load?
>
> **Null hypothesis:** In salmon, sex and parasite load are completely unrelated; knowing the sex of a salmon would not change your expectation of how many parasites it carries.
>
> **Question:** Do five different yeast strains differ in their growth response to temperature?
>
> **Null hypothesis:** The five different yeast strains do not differ in their temperature sensitivities.
>
> **Question:** Do athletes who have been allowed to rehydrate during a marathon run faster than those not able to rehydrate?
>
> **Null hypothesis:** There is no overall difference in running speed between athletes allowed to rehydrate and those unable to rehydrate.

Question: Do white blood cells from individuals who self-report themselves to be stressed differ in their activity from those from individuals who self-report themselves as not stressed?

Null hypothesis: There is no difference in the activity of white blood cells obtained from individuals who consider themselves stressed and individuals who consider themselves not stressed.

1.2.3 Why would we be interested in a null hypothesis?

Let us return to our multi-vitamin experiment introduced earlier, where we compare the school attendance of two randomly-selected sample groups of school children—one group who regularly takes a multi-vitamin, the other a placebo. Even if the null hypothesis were true, we would not expect our experiment to show that the mean number of days absent through ill-health in the two treatment groups were absolutely identical. We might, just by chance, have selected a few more children that are naturally prone to missing school through illness (regardless of their intake of vitamins) into one of the groups. Even if the children were absolutely identical in their likelihood of being absent from school, just by chance one child might contract a long-lasting illness and another might not. By analogy, two coins can be identical, but if we toss one we might get heads, but tails when we toss the other. Thus, we would not be completely surprised to find that the experiment showed a small difference between the groups of children. Imagine that in the experiment we found that the mean number of days missed by children in the placebo group was 14.2 and the mean number in the treatment group was 11.7, so that the difference in the means of the two groups was 2.5 days. The key question is whether this difference we observe in our samples is so small that it might be expected if the null hypothesis were true, or alternatively whether it is sufficiently big that it seems likely to have occurred only if the null hypothesis is not true. Null hypothesis statistical testing helps us make this decision.

In null hypothesis statistical testing, the data from an experiment are used to calculate some test statistic (some summary of the data), and along with that value of the test statistic the test provides a p-value. For our case, where we are comparing the means of two groups, a t-test might be a natural statistical test to use. If we apply this test to the data then the test provides us with a value of the t-statistic and an associated p-value. In our case, the p-value is the probability of an experiment like ours generating a difference between the two groups as big as (or bigger than) 2.5 days if the null hypothesis is in fact true and the treatment that an individual is assigned to has no effect whatsoever on the number of days of school missed.

More generally, we perform an experiment and apply an appropriate statistical test to the data from that experiment. Whether a given test is appropriate will depend on the question we are addressing, our understanding of the underlying biology of the system, and the design of our experiment. Applying the test to our data will generate a p-value. That p-value is the probability of obtaining a set of data at least as extreme as the data we have observed if the null hypothesis is actually true for the population that we are interested in. By 'extreme' we mean seemingly at odds with the null hypothesis. In our multi-vitamin example this is clearly a situation where there is a very big difference between the average number of school days missed between the two groups of children (regardless of which group produces the larger average number).

The p-value is often misinterpreted (and you can even find at least one example of a published paper by ourselves that fell into this trap), so the next section (1.2.4) describes in detail what a p-value isn't. You can see that if our experiment yields a p-value near one then the experiment gives very little support for doubting that the null hypothesis really is true; however, if the p-value is very small (much closer to zero than one) then this does give grounds for thinking that the null hypothesis might be quite unlikely to hold. The obvious question is how small is 'very small' in this context. This is an obvious question, but a good answer needs careful thought—so we will devote the section 1.3 to tackling this. Before we leave this section on the null hypothesis, we would like to dispel another common misconception. If your experiment yields a high p-value (that gives no reason to doubt that the null hypothesis holds), then this should not be seen as a disappointing result. In other words, null is not a synonym of dull (see section 1.2.5 if you require more convincing). If you feel your understanding of the concepts of this section would be strengthened by considering an example, then see section 1.2.6. Otherwise, if you feel you have a really good grasp of these concepts then you can safely move on to section 1.3 where we introduce the concept of errors associated with accepting or not accepting that the null hypothesis is true on the basis of your experiment.

1.2.4 What a p-value isn't

The p-value is the probability of an experiment like the one under consideration generating data at least as extreme as that actually seen if in fact the null hypothesis is true.

> The p-value is NOT the probability that the null hypothesis is not true.

> The p-value is NOT the probability that any other hypothesis is true.

So in the multi-vitamin experiment the p-value is *not* the probability that multi-vitamins have an effect on school attendance, it is *not* the probability that multi-vitamins have a positive effect on school attendance, and it is *not* the probability that multi-vitamin supplements reduce the number of school days missed by the observed difference of 2.5 days per year. If the p-value is so low that we consider that the null hypothesis is unlikely to hold, then logically this suggests that the experiment can be taken as evidence supporting the belief that multi-vitamins affect school attendance, but the p-value itself does not allow you to quantify how likely it is that this effect is real, or the direction and strength of that effect. You can make informal inferences as to the possible direction and strength of the effect from exploration of the results of the experiment, but the p-value does not contribute to that exercise. Formal description of the likely strength and direction of the effect takes us away from null hypothesis testing and into the realm of parameter estimation.

> A very small p-value need not imply that there is an effect of very large magnitude.

This seems a good place for us to define 'effect size', because this term will crop up throughout the book. We have already defined a parameter as a property of the underlying population that we are interested in. The effect size is the quantification of that parameter. For example, if the parameter we are interested in is the reduction in days off school associated with taking a daily multi-vitamin, the effect size associated with that might be three days if at a population level the average reduction was three days. An effect size of minus 1.2 days would imply that at a population level, taking a daily multi-vitamin would actually lead to 1.2 more days off school on average. The null hypothesis of no effect corresponds to an effect size of zero.

All other things being equal, if the null hypothesis is not true then the larger the real effect size the smaller the p-value will be. But all things are rarely equal, and lots of other aspects of the underlying population that you are investigating also affect the p-value, as do lots of aspects of the details of your experimental design. For example, if you carry out a really big study, you can obtain a very small p-value in a situation where the null hypothesis is false but the true effect size is only very, very slightly different from zero.

1.2.5 A null hypothesis is not a dull hypothesis

A null hypothesis is usually a description of no effect occurring, but sometimes no effect can be interesting. If you could demonstrate in a substantial and well-designed scientific experiment that any of the following null hypotheses seemed true then this would potentially be very interesting to a lot of people and have social and political implications:

- Moderate levels of alcohol consumption have no effect on driving ability
- A university education has no effect on lifetime earnings
- Growing foodstuffs organically has no impact on their nutritional value
- The death penalty has no deterrence effect on would-be criminals
- The cost of a pair of shoes is unrelated to how long they last

Just to avoid litigation—we are not offering any opinion as to the likelihood of these null hypotheses. However, if someone produced strong evidence in support of them then we doubt that their research would be considered dull.

Case study 1.2.6

Tossing a coin and p-values

Imagine a friend brought a coin to you and said they were concerned it was biased because they tossed it ten times and it turned up heads on nine of these occasions. We can consider their evidence in terms of a p-value. The null hypothesis in this case is that, independently of the outcome of any previous tosses, if you toss the coin it is likely to fall as heads with probability 0.5 and tails with probability 0.5. The p-value is the probability of obtaining data like that observed or more extreme under the null hypothesis.

If we toss an unbiased coin once, the probability of getting a head is 0.5. If we toss it twice then the probability of getting a head both times is 0.5 times 0.5, or 0.5^2. The probability under the null hypothesis of obtaining nine heads and one tail is simply $10(0.5)^{10} = 0.00976$. The factor of 10 occurs because the uncommon value might have occurred on the first toss, or the second, or the last toss. The probability of getting ten heads is $(0.5)^{10} = 0.000976$. So the probability of getting nine or ten heads is $0.000976 + 0.00976 = 0.011$. Now, the p-value is the probability of getting a result as extreme or more extreme than the observed value of nine heads and a single tail. We have just calculated the probability of getting nine or ten heads, but to get the p-value we also need to consider the probabilities of getting nine or ten tails, because these outcomes are just as extreme or more extreme as the observed outcome of nine heads. Hopefully you can see that for an unbiased coin the probability of getting either nine or ten tails is just the same as the probability of getting nine or ten heads. Thus we can finally calculate the p-value as $0.011 + 0.011 = 0.022$.

So you could report to your friend that the chance of observing their outcome of nine heads from ten tosses if the coin is truly unbiased is about 2.2 per cent. You have not been able to tell them definitely that the coin is biased or unbiased, but you have given some information that can help them make their best guess as to whether the coin is biased or not. You and your friend might conclude that it would be useful to subject the coin to more rigorous examination. You and she might wonder what range of outcomes you might expect if you tossed this coin 30 times and report the number of heads you obtain. Happily, the computer simulation approach that we champion in this book is ideal for exploring situations like this. At minimal effort we can simulate our experiment to see what range of outcomes you might expect and on the basis of this decide whether our experiment seems worthwhile carrying out in reality, or whether we might want to refine it first.

We just ask the computer to produce a list of 30 items, each of which has a 50 per cent chance of being heads and a 50 per cent chance of being tails. We then ask the computer to count up the number of heads in that list and remember that total. We then ask it to repeat that process for 100 lists to give us an idea of the range of possible outcomes. If we do that we get a distribution of values that looks like Fig. 1.2(a). We get a distribution because not all

Fig. 1.2 (a) The distribution of number of heads seen in 100 simulations of tossing an unbiased coin 30 times. (b) The distribution of associated p-values for the unbiased coin: 3 of the 100 simulations produce p-values less than or equal to 0.05. As we will discuss later, researchers often consider p-values below 0.05 as being small enough to support rejection of the null hypothesis. (c) The distribution of number of heads seen in 100 simulations of tossing a slightly biased coin (with a 55% chance of producing a head) 30 times. (d) The distribution of associated p-values for the slightly biased coin: 7 of the 100 simulations produce p-values less than or equal to 0.05.

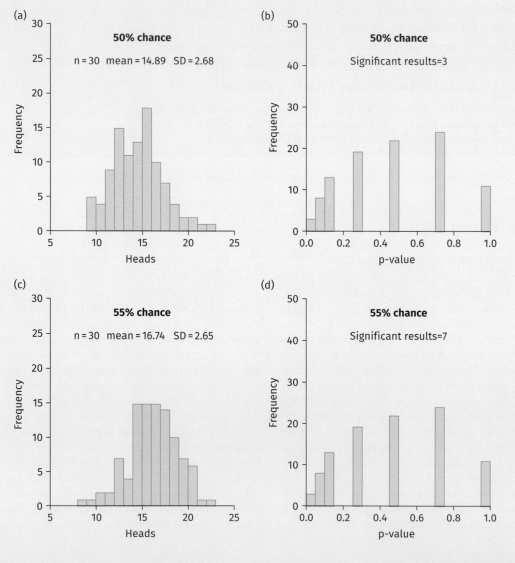

lists of 30 will contain the same number of heads; there is a bit of variation just by chance. But as we would expect, the values 15 and 16 are most common, and values further and further away from these values get progressively less likely.

For each list we could calculate an associated p-value, as shown in Fig. 1.2(b). We could ask the computer to repeat this procedure for a slightly biased coin that has a 55 per cent chance of giving a head at each toss. The results of this are shown in Figs 1.2(c) and 1.2(d). We see that this distribution of total number of heads is now biased a little towards higher values, and the p-values are generally lower. However, there is a lot of overlap between the distributions produced by an unbiased coin and a biased coin. This tells us that an experiment where we tossed the coin 30 times is not really going to allow us to differentiate between an unbiased and a slightly biased coin. Our null hypothesis is that the coin is unbiased. A small p-value might be taken as evidence to justify rejecting this null hypothesis (and

logically concluding that the coin is biased). As we will go on to discuss, a commonly used rule of thumb is that a p-value below 0.05 is 'small enough'. In statistical jargon a p-value below the critical value for being small enough is called a 'significant result'. We can see in Fig. 1.2 that our experiment would be expected to produce similar probabilities of a significant result (a p-value below 0.05) for the case of an unbiased coin (3 out of 100 simulations) and a slightly biased coin (7 out of 100). Most times when we carry out the experiment with a slightly biased coin we would not expect the experiment to provide sufficient evidence to reject the null hypothesis. We should probably think about changing our experiment to one that offered more divergent outcomes for the two coins.

Maybe tossing the coin 100 times might be more effective. It is trivial to ask the computer to simulate this bigger experiment, and the results are shown in Fig. 1.3. Now we have a slightly better chance of distinguishing between

Fig. 1.3 (a) The distribution of number of heads seen in 100 simulations of tossing an unbiased coin 100 times. (b) The distribution of associated p-values for the unbiased coin: 4 of the 100 simulations produce p-values less than or equal to 0.05. (c) The distribution of number of heads seen in 100 simulations of tossing a slightly biased coin (with a 55% chance of producing a head) 30 times. (d) The distribution of associated p-values for the slightly biased coin: 18 of the 100 simulations produce p-values less than or equal to 0.05.

(a)
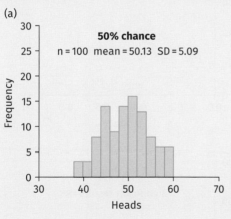

(b)
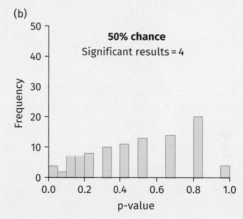

(c)
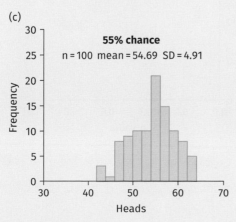

(d)
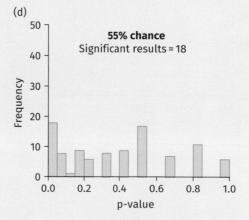

a slightly biased coin and an unbiased coin, but in truth our ability is still not strong (4 versus 18 significant results). Still, most times when we carry out the experiment with a slightly biased coin we would not expect the experiment to provide sufficient evidence to reject the null hypothesis.

Maybe we are going to have to bite the bullet and toss it a thousand times. You guessed it, the computer can show us how that might look in a flash (see Fig. 1.4). This looks a lot more promising (5 versus 88 significant results, so an experiment with a slightly biased coin is 9 times more likely to suggest rejection of the null hypothesis than an experiment with an unbiased one), and we would expect that the overwhelming majority of times when we carried out this experiment with a slightly biased coin, the outcome of the experiment would be in support of rejecting the null hypothesis. Before we embark on those 1000 tosses, we would be silly not to see how effective 500 tosses would be; that might save us some substantial thumb strain. ☺

Fig. 1.4 (a) The distribution of number of heads seen in 100 simulations of tossing an unbiased coin 1000 times. (b) The distribution of associated p-values for the unbiased coin: 5 of the 100 simulations produce p-values less than or equal to 0.05. (c) The distribution of number of heads seen in 100 simulations of tossing a slightly biased coin (with a 55% chance of producing a head) 1000 times. (d) The distribution of associated p-values for the slightly biased coin: 88 of the 100 simulations produce p-values less than or equal to 0.05.

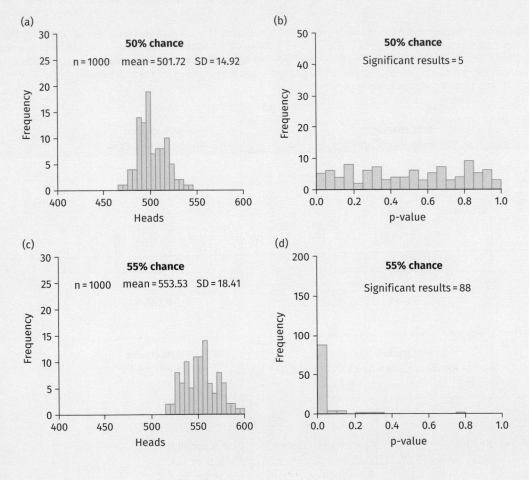

> ### 💡 Key point
>
> The p-value is the probability of an experiment like the one under consideration generating data at least as extreme as that actually seen if in fact the null hypothesis is true.

1.3   Type I and Type II errors

Earlier in this chapter, we defined the p-value as the probability of an experiment like the one under consideration generating data at least as extreme as that seen in the actual experiment if in fact the null hypothesis is true. We then said that if our experiment yields a p-value near 1, then the experiment gives very little support for doubting that the null hypothesis really is true; however, if the p-value is very small (much closer to zero than 1) then this does give grounds for thinking that the null hypothesis might be quite unlikely to hold. We also raised the obvious question: how small does a p-value need to be before we can reject the null hypothesis as untrue? The question might be obvious, but trying to provide a satisfactory answer is a huge challenge, and gets to the heart of why some researchers consider null hypothesis statistical testing to be unsatisfactory. (This isn't the place for us to critique different statistical techniques, but if you want to read more about strengths and weaknesses of null hypothesis statistical testing, then Stephens et al. (2005) might be a good place to start.)

One answer to the how-small question could be that the question is logically flawed. We can never reject the null hypothesis with certainty. Remember that we began this chapter by explaining that rarely can an experiment based on a small sample definitively answer questions about the wider population from which the sample was drawn (see the Electronic Supplementary Material for this chapter on the associated website for discussion of rare exceptions). Thus, the answer 'the question is logically flawed' is formally correct, and definitely worth remembering, but it is not really of much practical value. It should, though, cause us to rephrase the question to: how small does the p-value have to be before we feel that (although we cannot know for certain) it seems reasonable to act on the assumption that the null hypothesis is false?

To see how we might answer this question in a more practical way, we are going to introduce two statistical terms: Type I and Type II errors. For any study, there are four possible outcomes, and these are shown in Table 1.2. The rows in the table represent the two possible states of the world, the null hypothesis is either true, or it is false … it has to be one or other. This depends on the underlying biology. Of course, the researcher does not know the true state of the world when they carry out their study, otherwise they would not need to perform the study in the first place. However, based on their data, their statistical analysis, and ultimately the p-value they obtain, they will make a decision about whether they have enough evidence to reject the null hypothesis or not. So the two columns in the table represent the two options the researcher has, based on their data.

Table 1.2 The four possible combinations of the true validity of the null hypothesis at population level and the conclusion about it drawn from a sample.

Reality/inference from sample	Rejection of null hypothesis on basis of sample	Not rejecting null hypothesis on basis of sample
Null hypothesis is true	Type I error (false positive)	Correct inference
Null hypothesis is false	Correct inference	Type II error (false negative)

Now let's focus on the top row in the table, the situation where the null hypothesis really is true. If, based on their p-value the researcher decides not to reject the null hypothesis, then they end up in the top-right box of the table and have got the correct answer. There is nothing going on, and they correctly conclude that. If instead they choose to reject the null hypothesis based on their p-value, then they end up in the top-left box and have got the wrong answer. They have incorrectly concluded that something is going on, when the reality is that it is not. Statisticians call this kind of false positive result a Type I error. As it turns out, the probability of ending up in one or other box in the top row of the table depends simply on how small the researcher decides a p-value must be, before they are willing to reject the null hypothesis. So imagine our researcher decides that if their p-value is 0.1 or less, they are willing to accept that as strong enough evidence to reject the null hypothesis. If the null hypothesis really is true (and this is a big and important 'if' that we will come back to a little later), then their probability of being in the Type I error box rather than the correct box is 0.1. Put another way, they would conclude something was going on when it wasn't 10 per cent of the time. Now being wrong 10 per cent of the time might seem quite high, so perhaps our researcher decides that actually a smaller p-value threshold would be better. If they require a p-value of 0.05 before they reject the null hypothesis, then they will make a Type I error 5 per cent of the time or, for a p-value threshold of 0.01, only 1 per cent of the time. So deciding how small the p-value needs to be before you will reject the null hypothesis comes down to deciding what risk of Type I errors you are comfortable with.

Now scientists don't like making errors of any kind, so an obvious question that you might be asking yourself right about now, is why not make the p-value threshold really really really small—say 0.00001, or 0.00000001? That way we would be really unlikely to make a Type I error. This is a good point, but it overlooks one important issue. Up until now we have always been assuming the null hypothesis is true. So let's think about the other scenario, when the researcher is on the bottom row of our table of outcomes and the null hypothesis really is false. As before, our researcher has to make a decision as to whether to reject the null hypothesis based on the p-value from their experiment. If their p-value falls below their desired threshold, then they reject the null hypothesis and end up in the bottom-left box, drawing the correct conclusion. However, if their p-value is above their chosen threshold then they end up in the bottom-right box, and draw the wrong conclusion. This kind of mistake is called a Type II error by statisticians, and it is a false negative. The difference observed in the data is not extreme enough to meet our desired level of proof. As above, the probability of being in the Type II error box depends on the critical p-value threshold we choose for our study. The situation is not quite as straightforward as with the Type I error, as there is no simple correspondence between the size of the probability of making a Type II error and the p-value threshold. However, there is a rule that is always true, for any given study: the smaller the critical p-value threshold, the higher the chance that the researcher will make a Type II error. This makes intuitive sense; by reducing the size of the p-value we want to see before we are happy to reject the null hypothesis, we are asking for a greater level of proof. If the null hypothesis is false, then this need for greater proof means we are less likely to erroneously conclude that something is going on when it is not, but it means our data must also meet more stringent standard of proof before we can correctly reject a false null hypothesis. In the extreme, we might set our threshold so low as to effectively mean we would never make a Type I error, but the cost of doing so would be that we would never be able to detect real effects.

All of this is a very long way of saying that the answer to our question 'How small does the p-value have to be before it seems reasonable to act on the assumption that the null hypothesis is false?' depends on a balance between the risk of making a Type I error we are willing to tolerate, and the consequences on our risk of making a Type II error. This is not a question that has a single correct answer, and any answer will ultimately be arbitrary, but that is not to say that it is not important to think about, and understand the consequences of different answers. And also to keep in mind that whatever threshold we choose, this transparency and decisiveness comes at a cost of sometimes being wrong.

Having said all of that, there are some conventions (arbitrary conventions certainly, but conventions nonetheless): a large body of practising researchers have all adopted the convention that p-values less than 0.05 are sufficient grounds to reject a null hypothesis and p-values equal to or greater than 0.05 are not. Let us be quite clear—this threshold has no justification other than convention. Indeed, a threshold set at any value has no justification other than convention. In reality, there are no logical grounds for treating a p-value of 0.049 as different from 0.051, but adopting the convention means that you must do that as the necessary price to be paid for adopting a clear and simple criterion for decision-making on the basis of a p-value.

If you buy into the idea that we can adopt this convention to base our conclusions on where our measured p-value sits relative to an arbitrary threshold in order to let us make clear decisions, then null hypothesis testing is for you. If you philosophically cannot accept this, then you need to pursue other statistical approaches. For the next few chapters we are going to assume that you are on board with this convention. Before we move on, we note that there are times when researchers adopt critical values other than 0.05 (although in the life sciences 0.05 is very much the most common choice) and we will discuss reasons for this in Chapter 3.

1.4   How do we define statistical power?

Under null hypothesis statistical testing, statistical power can be defined very simply:

> The statistical power of an experiment is one minus the likelihood of making a Type II error on the basis of the outcome of that experiment.

To put it another way:

> The statistical power of an experiment is the probability that you will correctly reject the null hypothesis if in fact the null hypothesis is untrue for the population of interest.

Now you can see why we took you through all the preliminaries for most of this chapter—these were the foundations for giving you the clearest possible understanding of what statistical power is. With those foundations firmly in place, defining statistical power is very easy. Clearly, you can see that statistical power is a probability, and so will range between 0 and 1. We can see from this definition that high statistical power is a desirable feature of an experiment. This book aims to help you understand the factors that influence statistical power (Chapter 3), and calculate the power of any potential experiment you might be thinking of performing (Chapter 4), and then on the basis of these calculations decide and justify on the basis of power any experiments that you actually propose to do (Chapters 5–8). Before we do that, we want to spend the next chapter emphasizing that low statistical power is highly undesirable. This may seem obvious, but in fact the reasons why performing a low-powered experiment is undesirable are more subtle and interesting than many practising scientists realized until the last decade or so.

⪦ Summary Points

- We can almost never study the whole population that we are interested in; instead, we study a representative sample.
- Different forms of inferential statistics are complementary; neither null hypothesis statistical testing, parameter estimation, nor model selection is universally superior to the others.
- The p-value obtained from the data generated in an experiment is the probability of observing those data (or even more extreme data) if the null hypothesis were actually true.
- Using a p-value in a decision to accept or reject the null hypothesis as true runs the risk of making Type I and Type II errors
- Type I and Type II errors are related, such that you cannot simultaneously minimize the risk of occurrence of both.
- The statistical power of an experiment is the probability that you will correctly reject the null hypothesis if in fact the null hypothesis is untrue for the population of interest.
- That is, the statistical power of an experiment is one minus the likelihood of making a Type II error on the basis of the outcome of that experiment.

⪦ Further reading

Goodman, S., 2008. A dirty dozen: twelve p-value misconceptions. *Seminars in Hematology*, 45(3), pp.135–140. W.B. Saunders. [A comprehensive discussion of what a p-value is and isn't.]

Greenland, S., Senn, S.J., Rothman, K.J., Carlin, J.B., Poole, C., Goodman, S.N., and Altman, D.G., 2016. Statistical tests, P values, confidence intervals, and power: a guide to misinterpretations. *European Journal of Epidemiology*, 31(4), pp.337–350. [A detailed examination of all the issues raised in this chapter.]

Lieberman, M.D. and Cunningham, W.A., 2009. Type I and Type II error concerns in fMRI research: re-balancing the scale. *Social Cognitive and Affective Neuroscience*, 4(4), pp.423–428. [A thoughtful discussion on getting the right balance between Type I and Type II errors.]

Rock, C.L., 2007. Multi-vitamin-multimineral supplements: who uses them? *The American Journal of Clinical Nutrition*, 85(1), pp.277S–279S.

Ruxton, G. and Colegrave, N., 2011. *Experimental Design for the Life Sciences*. Oxford University Press. [Selecting a sample that is truly representative of the population of interest is one of the bedrocks of good experimental design—and we discuss ways to do this (and pitfalls to avoid) here.]

Stephens, P.A., Buskirk, S.W., Hayward, G.D., and Del Rio, C.M., 2005. Information theory and hypothesis testing: a call for pluralism. *Journal of Applied Ecology*, 42(1), pp.4–12. [A discussion of different types of inferential statistics, written by biologists for biologists.]

Discussion questions

1.1   Can you think of reasons why using number of days missing from school through ill-health might not be the ideal way to measure a child's health?

1.2   If you were modelling factors that influence children's non-attendance at school through ill-health what factors would you expect to consider in your model?

1.3   The best balance between the risks of making Type I and Type II errors is going to depend on the consequences of the two types of error and the probability that the null hypothesis is actually true. Discuss that balance in the context of: (a) drug discovery (i.e. testing compounds for their potential in medicine); (b) pre-flight safety checks on an aeroplane; and (c) treating someone with an anti-venom that has very nasty side-effects because they appear to have been bitten by a highly venomous snake.

1.4   If you tossed a coin 100 times, what is your instinct about how far the ratio of heads and tails would have to deviate from 50:50 for you to be concerned that the coin might be biased?

2 WHY LOW POWER IS UNDESIRABLE

Learning objectives

By the end of the chapter, you should be able to explain:

- Why a low-powered study has a high probability of failing to detect genuine effects.
- How low power makes drawing useful inferences from failure to reject the null hypothesis more difficult.
- Why, if an experiment is low-powered, the risk that a significant result is actually a Type I error is increased.
- Why low power inflates the effect sizes associated with statistical significance.

At the end of the last chapter we defined statistical power:

The statistical power of an experiment is one minus the probability of making a Type II error based on the outcome of that experiment. If the null hypothesis really is false, it is the probability that we will correctly reject the null hypothesis.

It seems almost obvious then that low power is undesirable—because low power means increased risk of a Type II error. Errors are bad; therefore low power must be bad. Thus, at first glance, this chapter might seem a waste of your time. However, in fact a string of highly influential recent publications has changed how we think about low power—and made us appreciate that it is much more undesirable than many of us realized. We will unpack these ideas in this chapter. But we will still strive to keep this chapter as short as we can—since the take-home message is simple: you should try to avoid carrying out low-powered studies yourself, and you should be very cautious about drawing inferences from anyone else's study that seems low-powered to you. In the remaining sections of this chapter we will explain exactly why.

2.1   With a low-powered study you risk missing interesting effects that are really there

Imagine in a comparative study you find that there are 20 different compounds that seem to be more common in seeds with long dormancy periods compared to those with shorter dormancies. You hypothesize that some of these compounds might have antibiotic properties that protect the seed from infection during its extended period of dormancy. You design an assay to test each of these compounds individually in terms of their ability to inhibit the growth of bacteria. Unknown to you, four of these 20 compounds are actually effective antibiotics.

Let us first of all assume that you carry out the same experiment on each compound in turn (for the sake of this example, we will ignore the fact that carrying out separate experiments for each may not be the most efficient way to proceed). Each experiment is well replicated, and (because of this) you spend an awful lot of time measuring bacterial growth. However, your reward for all of this hard work is that your experiments will have high power. Let us assume that the power of each experiment is 0.8 (which, as we will discuss in the next chapter, is generally considered high power). Then your probability of correctly detecting all four antibiotic compounds is simply 0.8^4 which equals 0.41. Your probability of detecting at least three of them is 0.82 $(0.41+4*0.2(0.8)^3)$, and your probability of detecting at least two of them is 0.97 $(1-(0.2)^4-4*0.8*(0.2)^3)$. The probability of you missing all four is only 0.0016 $((0.2)^4)$.

Now suppose instead that you could not face doing all of those bacterial assays, and so decided to save time and money by measuring fewer replicates. Let's assume that the consequence is that your power drops from 0.8 to 0.2. Now the probability of you hitting the jackpot and detecting all four antibiotic compounds is only 0.0016 (0.2^4), which is a lot less than the probability of you sharing your birthday with your mother; whilst your probability of missing all four is 0.41, which is really not far off the odds of flipping a coin and getting heads.

This example encapsulates the most obvious risk associated with low power—failing to detect interesting effects. In this case, you had a really creative and exciting idea to look for compounds that were particularly associated with long seed dormancy periods, and hypothesized that some of these compounds might have antibiotic properties. It turns out that your idea wasn't just good—it was correct. In fact, four of the 20 compounds that you identified were antibiotics. If you follow up your good idea by screening each compound in a high-powered study then you are almost certain to discover at least two of these and maybe all four. However, if you used a low-powered study instead, then there is a very real chance (41 per cent) that you would end up failing to discover the antibiotic properties of any these four compounds.

The decision to go with a high-powered test could lead to you revolutionizing human and animal healthcare. The alternative decision to go with a lower-powered test might mean that you wrongly conclude that your seed-compound idea wasn't so good after all, and (even worse) telling others that you checked out seed compounds associated with long dormancy but they didn't seem promising as a potential source of antibiotics. A huge opportunity could be missed. If you are anything like us, then you don't get really good ideas all that often. In our scenario here you had been blessed with a great idea—but that could go entirely unrecognized through carrying out low-powered experiments.

Key point

The first and most straightforward drawback to a low-powered study is that it has a high probability of failing to detect (interesting) effects.

2.2 You cannot read much into lack of statistical significance in a low-powered study

As we discussed in the last chapter, failure to reject a null hypothesis should not automatically be seen as a disappointing outcome. Sometimes failure to reject the null hypothesis can be interesting and/or helpful to a line of research. For example, imagine a study of a new painkilling drug fails to reject the null hypothesis that the drug has no side-effects in terms of reduced concentration and mental sharpness. This can be seen as an encouraging result that might indicate that the drug is more widely useful and more acceptable to potential users. However, how interesting this study is rests very much on its power. If a high-powered study found no effect, then you could infer that if there had been a substantial effect of the drug to find then the study would very likely have found it. Thus it does seem that the findings of such a study are really useful in terms of evaluating the potential utility of this new drug. If, however, it is a low-powered study, then a very real possibility is that actually, in the situation under study, the drug did significantly impair concentration but the study was quite likely to miss such an effect. Thus, you erroneously suggest that the drug has no side-effects (making a Type II error).

 Key point

In a low-powered study that fails to find an effect that you were interested in, it might be that the effect really was not there, but it might also very well be that the effect was there but simply went undetected.

2.3 You cannot read much into statistical significance in a low-powered study

It's just possible that you might be thinking that one implication of the previous sections is that if you did obtain a significant result despite having low power then this would be a very strong indication indeed that there really was a real underlying effect. Sadly, the opposite is true; the lower your power, the less likely it is that a significant result is actually triggered by a real underlying effect.

To see why, let's work through a simple numerical example (for a more thorough treatment of this issue, see section 2.3.1). Imagine that around the globe there are 200 different research labs, each investigating a different treatment for cancer. For simplicity, let's assume that 100 of these treatments are effective against cancer (i.e. the null hypothesis is false), and in the other 100 labs the treatments under investigation have no effect at all (i.e. the null hypothesis is true). Of course, none of the researchers know, for their particular treatment, whether the null hypothesis is true or false. Each group then carries out a study, collecting and analysing the data. Then, based on the outcome of their statistical test, each decides whether to accept or reject the null hypothesis that their treatment has no effect against cancer. In common with many areas of biology, they base this decision on whether their p-value is less than or equal to 0.05 (in statistical speak, they accept a Type I error rate for their study of 5 per cent).

Let's first focus on the 100 labs where the treatments are effective. Based on their experiment, the researchers have two possibilities: they either get a significant result in their statistical test, reject the null hypothesis and draw the correct conclusion (the treatment is effective), or they fail to reject the null hypothesis, even though it is actually false, and they make a Type II error. Hopefully you can see that the probability of each of these outcomes depends on the power of their studies. Let's assume that these are all well-funded labs with the resources that allow them to carry out high-powered experiments. If the power is 0.8, then we expect about 80 of these 100 labs to obtain statistically significant results, whilst 20 get non-significant results. This situation is illustrated in the first panel of Fig. 2.1.

Now let's change our focus to the 100 labs unknowingly exploring ineffective treatments. If their statistical test produces a non-significant result, they will fail to reject the null hypothesis (and consequently get the correct answer from their study). However, if the test produces a significant result, they will falsely reject the null hypothesis and make a Type I error. Since we are

Fig. 2.1 A hypothetical example where 200 laboratories are each testing different cancer treatments, exactly 50% of which are effective. We compare the 100 labs where the treatment has a real effect and the 100 labs where it does not, in two scenarios (power to detect a real effect is either 80% or 20%).

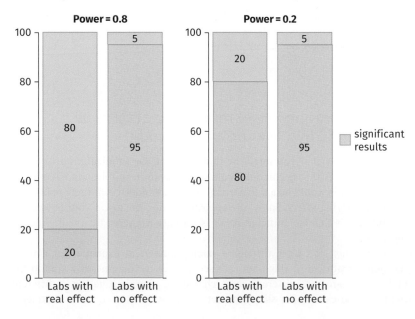

assuming a Type I error rate of 5 per cent, then this means we expect about 95 of these labs to get the correct answer, and about five to get spuriously significant results.

Across all 200 labs, 85 sets of researchers reported that their treatment was effective based on a statistically significant result. In 80 of these cases, the researchers were right, and so the vast majority of the positive results reported (about 94 per cent) are real. That is, 85 treatments are reported as effective against cancer, and 80 of these treatments actually are.

But what would have happened if the studies had had lower power? This scenario is shown in the second panel of Fig. 2.1. The situation in the labs looking at ineffective treatments stays the same because the Type I error rate is unaffected by power. So these labs produce about 5 false positives, as before. However, the situation with the other 100 labs is very different. If their studies have a power of 0.2, this means we now expect only 20 of the labs to correctly report effects based on significant results. In total, our 200 labs have only reported 25 positive results, and only 80 per cent of them are true positives. The reduction in power means that the proportion of positive results we get that are false increases (from 6 per cent with high power up to 20 per cent with low power).

Now this may not seem too bad to you. Sure, the proportion of false positives has increased, but it is still the case that most of the times we appear to detect something, it is a real effect. However, we have assumed we are working in a research field where the null hypothesis is often false (remember that 50 per cent of the treatments had an effect). Suppose we are working in a more exploratory field of biology, where real effects are far rarer, and the null hypothesis is usually true. Suppose that rather than the null hypothesis being false 50 per cent of the time, it is false 10 per cent of the time. This would mean that only 20 of our 200 labs are now testing effective treatments, whilst in the other 180 labs the null hypothesis is true. In this situation, if our studies only have a power of 0.2, we would only expect about four of these 20 labs to report significant results, whilst we would expect about nine false positives from the other 180 labs. That means that in our global study, about 69 per cent of the reported positive effects are false. In a real research field, that could equate to an awful lot of time and money being spent following false leads. In this case, with 0.8 power we would expect 16 true positives and still nine false positives, so the fraction of false positives would drop from 69 per cent when power is 0.2 to 36 per cent when power is 0.8. To us, this seems like a much more bearable situation than the low-powered case.

Now put yourself in the position of one of these cancer treatment researchers. Your study has just produced a significant result and you have rejected the null hypothesis. If your study had high statistical power, then the probability that you have discovered something real is relatively high. You can be confident about sharing it with the world and planning the next phase of your study based on this result. However, if your study had low power, and especially if you work in an exploratory field of science, you can have little confidence that your significant p-value indicates something real rather than simply being a Type I error.

The ramifications of the arguments presented in this section are pretty seismic. Firstly, there may be good reason to abandon an investigation if practical constraints mean that you have low power, because no matter whether the final statistical test in a low-powered investigation suggests an underlying effect or not, you cannot put much faith in this.

Secondly, you cannot trust the results of someone else's investigation, if that procedure was low-powered. In a hugely influential paper in 2005, John Ioannides presented the ideas summarized here along with arguments for why we might expect that most published studies actually have rather low power—this allowed him to give his paper the very arresting title 'Why most research findings are false'. As an aside, this publication is part of the modern trend towards open access, where anyone can have access to the full paper without paying. Put this to the test by typing in the name of this paper into your favourite search engine, and you should be able to download a full version of the paper without difficulty. You might be surprised that low-powered studies are published, given what we have discussed in this chapter so far. Actually, it seems that low power is the norm: investigations across different fields of research have come back with estimated powers in published studies between about 20 per cent and 50 per cent.

If you feel you have a good understanding of how low power can make interpretation of rejection of the null hypothesis more difficult, then it is safe to go on to section 2.4 of this chapter, but if you would like to solidify the ideas introduced here through a more formal treatment, then try section 2.3.1.

2.3.1 The positive predictive value as a useful concept in interpreting the results of an experiment

Consider the following scenario. You perform a number of experiments over time, testing different hypotheses. Let us assume that in one in five of these experiments the null hypothesis is actually true, and in the other four times out of five the null hypothesis is actually false. It is conventional in statistical theory to define something called the pre-study odds (denoted R). This is the odds (before we collect any data in our focal experiment—hence 'pre-study') that the null hypothesis in a given experiment is actually true. Thus, R is the fraction of tests with an underlying real effect divided by the fraction of tests with no underlying effect. In our example case, the value of R is ($\frac{1}{5}$) divided by ($\frac{4}{5}$), which simplifies to $\frac{1}{4}$.

What we are really interested in is the probability that, if our statistical test does suggest that there is an effect, this effect actually exists. This is traditionally called the positive predictive value (denoted PPV). We can actually calculate this from simple probability theory if we know the value of R, the power of our study ($1 - \beta$, where β is the probability of making a Type II error)—and the Type I error rate that we set for our statistical test (α).

The probability of there actually being a real effect is $R/(R + 1)$. In such a situation, the probability of detecting this effect is $(1 - \beta)$—remember this is the definition of power. So the probability of an experiment yielding a positive statistical result because there is a real effect is $R \times (1 - \beta)/(R + 1)$. Let's call this P_1.

The probability of there actually not being a real effect is $1/(R + 1)$. So the probability of your statistical test suggesting an effect when there is not actually one (call this P_2) is simply $\alpha/(R + 1)$.

Remember that we want to get PPV, which is the probability that a statistical test suggesting an effect is caused by there actually being an effect. This is simply $P_1/(P_1 + P_2)$. After a little bit of algebraic rearrangement this gives us:

$$PPV = \frac{R(1 - \beta)}{R(1 - \beta) + \alpha}.$$

Fig. 2.2 As the power of a study increases, so does the probability of a statistical test suggesting a result being caused by there actually being an effect (PPV). This is true across a range of values for power and the pre-study odds (R). Where the Type I error rate for the statistical test (α) is as low as 0.01, PPV is greater than where $\alpha = 0.05$, but lower power still reduces the PPV value.

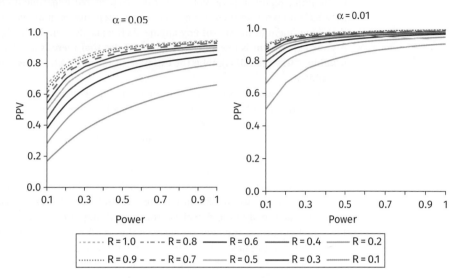

Let us assume that R is still 0.25 and α takes the traditional value 0.05. Then if the power $(1 - \beta)$ is 0.8 then PPV is 0.8, but if the power is 0.2 then PPV is 0.5. So, in this situation, if our statistical test suggests that there is an effect then this is in fact correct 80 per cent of the time if we have 80 per cent power, but falls away to only 50 per cent of the time if we had a lower power of 20 per cent.

Some simple calculus would show you that regardless of the values of R and α, the lower the power, the lower the PPV; thus the lower the power, the less you can rely on a statistical test yielding a significant p-value as a good indicator that there really is an underlying effect.

In Fig. 2.2 we illustrate this for a range of different values for power and R, both for $\alpha = 0.05$ and 0.01. Reducing the value of α does boost the PPV value but the simple fact remains that lower power always reduces the PPV value.

In practice we can never know the value of R, but lower power reduces PPV for all values of R. However, we can say that R will be lower in more speculative fields of enquiry like drug discovery, where hundreds of compounds might be screened in the hope of finding a handful that have any effect. Conversely, R might be a lot higher in clinical trials, where a drug will likely have already proven highly effective in laboratory studies and perhaps also in animal studies before the clinical trial is sanctioned.

 Key point

If an experiment is low-powered, then a p-value suggesting rejection of the null hypothesis is more likely to be a Type I error than if the experiment were high-powered.

2.4   Your estimation of the size of an apparent effect can be unreliable in low-powered studies

It shouldn't really surprise you that if a low-powered experiment cannot reliably tell you whether there is an effect at all, then any estimate you make from that experiment of the size of any apparent effect is also unreliable. Specifically, if your experiment is low-powered but indicates that there is an effect (e.g. the p-value is less than 0.05), then the estimate of the size of the effect will tend to be inflated.

One way to see this is to remember that when the experiment is low-powered, a higher fraction of positive statistical results (i.e. those suggesting an effect) are actually Type I errors. In such situations, a Type I error has been caused because, just by chance, your sampling has produced values indicative of an effect; now the actual effect is non-existent so the effect size associated with all these Type I errors is inflated.

However, imagine there is actually a small effect but the experiment is still low-powered. Because of this low power we will have a high probability that the experiment will fail to detect the real underlying effect. When it does reject the null hypothesis, this will be because the small real effect and the chance effects of sampling have worked in the same direction, producing what appears to be a larger effect. It is these apparently large effects that trigger the statistical test to return a low p-value. Thus, once again, if we are only interested in effect sizes when the statistical test suggests that there is actually an effect, then the size of the effects that we look at will tend to be inflated. This phenomenon of inflation is often called the winner's curse in the statistical literature. The arguments above might seem a little abstract, so it is probably useful to cement them with an example.

Case study 2.4.1

A hypothetical example of the winner's curse

Let us assume that we want to compare the IQ levels of left-handed and right-handed individuals. We do this by administering an IQ test to a number of individuals and then asking each of them whether they consider themselves to be predominantly left- or right-handed.

Let us first of all consider a hypothetical true underlying situation where the distribution of IQ scores is exactly the same in the populations of left-handers and right-handers, and that distribution is a normal distribution with a mean of 100 and a standard deviation of 15. The distribution of a thousand individuals' IQs is as shown in Fig. 2.3a. We should pause just to mention some properties of the normal distribution, since it is commonly encountered. A normal distribution has a symmetrical shape often compared to a bell. It is completely defined by the values of its mean and standard deviation. It is symmetrical about the mean and most values of a normal distribution are quite close to the mean. How close they are is determined by the standard deviation; the smaller this is, the closer most values are to the mean. A handy rule of thumb is that for a normal distribution, approximately 95 per cent of values lie within two standard deviations of the mean. So in our case in Fig. 2.3(a), since we have a large sample, we would expect that about 950 of our 1000 values will lie between 70 (100 − (2*15)) and 130 (100 + (2*15)).

Let us assume that for practical reasons we could only sample five left-handers at random from the population and five right-handers. Such small samples will reduce the power of our study. Imagine, further, that 1000 scientists around the world repeated this study (we know this is wildly improbable), and each of them measured the IQs of the two groups of five and used a Student's t-test to look for evidence of an effect of handedness on IQ. The distribution of the 1000 p-values we got when we simulated this situation is shown in Fig. 2.3(b). The most natural way to show the estimated effect size in each of the 1000 cases is to plot the mean of the five right-handers' IQs minus the mean of the five left-handers' IQ. These values are shown in Fig. 2.3(c). Since there is no underlying effect, the experiment is just as likely to throw up a negative or a positive difference. Most values are close to zero but some are larger.

Of these 1000 t-tests, 40 produced p-values less than 0.05 (which might be seen as suggestive of an underlying effect). This value of 40 out of 1000 seems plausible since we have set a Type I error rate at 0.05, so we should expect around 50 (0.05*1000) Type I errors. When we plot the effect sizes associated with these 40 replicates of the experiment in Fig. 2.3(d), then we see that these are the p-values associated with on average larger effect sizes. So, if you just carried out one of these low-powered experiments, and it yielded a significant p-value, then in this instance where handedness is not linked to IQ, this has to be a false positive. If you then investigated the difference between the two means to see the size of the apparent effect, then this is just as likely to suggest that left-handers have lower or higher IQs but the estimated difference on average would be about 21.

Fig. 2.3 (a) The distribution of 1000 individuals' IQs. (b) The distribution of p-values resulting from Student's t-tests on 5 left-handers' and 5 right-handers' IQs in 1000 studies. (c) The estimated effect sizes in these 1000 experiments, calculated by subtracting the mean of the 5 left-handers' IQs from the mean of the 5 right-handers' IQs in each study. (d) The estimated effect sizes in 40 of the 1000 studies, wherein the resulting p-value was <0.05 (mean of absolute values = ~21).

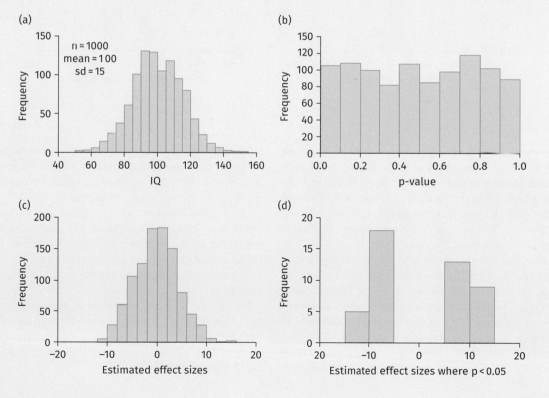

Fig. 2.4 (a) the distribution of 1000 individuals' IQs. (b) The distribution of p-values resulting from Student's t-tests on 25 left-handers' and 25 right-handers' IQs in 1000 studies. (c) The estimated effect sizes in these 1000 experiments, calculated by subtracting the mean of the 25 left-handers' IQs from the mean of the 25 right-handers' IQs in each study. (d) The estimated effect sizes in 41 of the 1000 studies, wherein the resulting p-value was <0.05 (mean of absolute values = ~9).

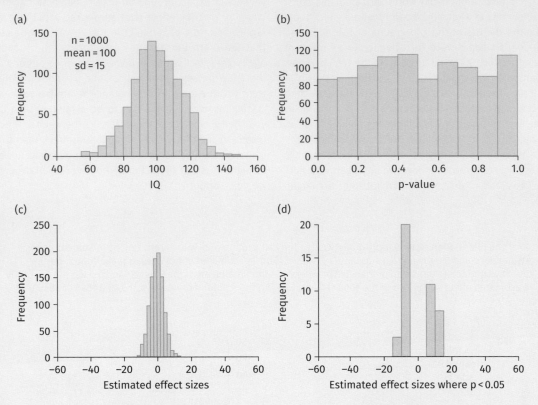

Figure 2.4 shows analogous results to Fig. 2.3, but now we are using sample sizes of 25 rather than five in each group to improve our statistical power. Notice that the probability of generating a false positive remains around 5 per cent (this time we generated 41 false positives), and if there is a false positive then the estimated effect size is reduced from around 21 to ten.

As a final-thought experiment, imagine that left-handers do have higher IQs on average than right-handers. Let's say that the mean for left-handers is 105 and for right-handers is 95, but the standard deviation is 15 in each case. Figure 2.5a shows the two distributions; the difference in means is of course ten. Now repeat our imaginary scenario where 1000 scientists around the world perform the same low-powered experiment comparing 5 left-handers and

5 right-handers. The results are shown in the remaining panels of Fig. 2.5: we see that we correctly detect that there is an effect in 143 cases, but when we look at the average effect size for these cases, it is inflated to 23.

Figure 2.6 is analogous to Fig. 2.5, but now we have higher power conferred by a 25 versus 25 experiment. We see first of all that we detect an effect more readily (in 645 cases), but also that the average estimated effect size from these (around 12) is closer to the correct underlying value of 10.

Thus, to summarize, if a low-powered experiment suggests that there is an underlying effect, then the data from that experiment will also tend to inflate the size of that effect. As with the last section, the ramifications of this can be substantial.

Fig. 2.5 (a) The distributions of 1000 left-handed individuals' IQs (mean = 105, SD = 15) and 1000 right-handed individuals' IQs (mean = 95, SD = 15). (b) The distribution of p-values resulting from Student's t-tests on 5 left-handers' and 5 right-handers' IQs in 1000 studies. (c) The estimated effect sizes in these 1000 experiments, calculated by subtracting the mean of the 5 left-handers' IQs from the mean of the 5 right-handers' IQs in each study. (d) The estimated effect sizes in 143 of the 1000 studies, wherein the resulting p-value was <0.05 (mean of absolute values = ~23).

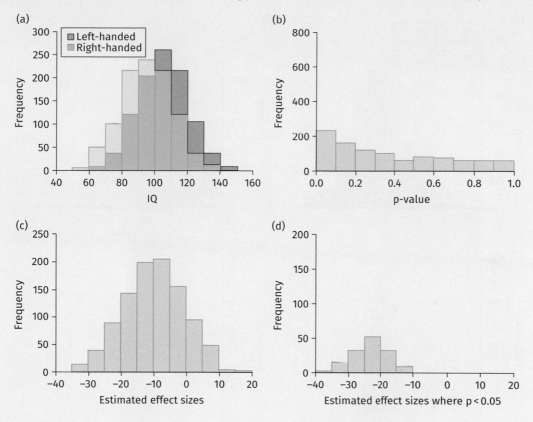

2.4.2 Ramifications of low p-values being associated with inflated effect sizes in low-powered experiments

The key ramification is the Proteus phenomenon, where a discovery of an apparently substantial and interesting effect is not backed up in subsequent studies that find less strong effects or no evidence of an effect. The likely explanation here is that the initial study was a speculative exercise not specifically designed to have high power for evaluating this effect, and has fallen foul of the phenomena described here. In contrast, the follow-up studies were purpose-designed to investigate this specific effect with high power and avoid these same pitfalls. These follow-up studies can be seen in a positive light as part of the self-correcting process of scientific development. A more scathing view is that the follow-up studies were so much wasted time and effort that could have been utilized more productively if more attention had been paid to evaluating the power of the initial study and then interpreting the results of the study in that light.

Fig. 2.6 (a) The distributions of 1000 left-handed individuals' IQs (mean = 105, SD = 15) and 1000 right-handed individuals' IQs (mean = 95, SD = 15). (b) The distribution of p-values resulting from Student's t-tests on 25 left-handers' and 25 right-handers' IQs in 1000 studies. (c) The estimated effect sizes in these 1000 experiments, calculated by subtracting the mean of the 25 left-handers' IQs from the mean of the 25 right-handers' IQs in each study. (d) The estimated effect sizes in 645 of the 1000 studies, wherein the resulting p-value was <0.05 (mean of absolute values = ~12).

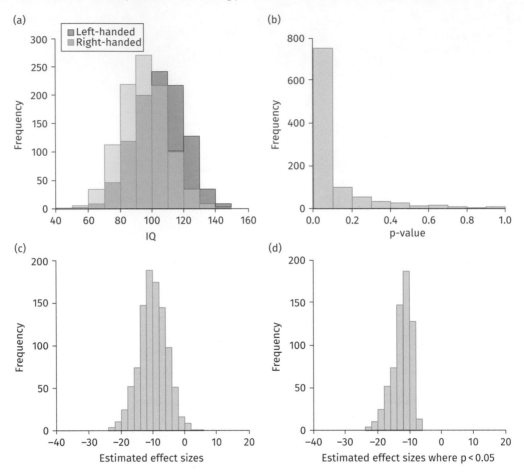

A more subtle ramification is that there is a danger that the follow-up studies end up being less powerful than the researchers intend. As we will discuss in the next chapter, one thing that must be stipulated when calculating power is the effect size that is of interest. The larger the effect size of interest, the higher the power. It would be natural to use the effect size of the initial study to influence the focal effect size in the follow-up studies, but that effect size is likely to be inflated compared to the real underlying effect size if the initial study was low-powered.

2.4.3 Don't let your belief in the worth of your experiment be over-influenced by the p-value it produces

The last point we want to make is that you will notice in our hypothetical IQ experiments that there can be quite a spread of p-values in our thousand replicate experiments of exactly the same phenomenon. This effect is particularly strong for low-powered experiments; however, power needs to be very high in order to really produce high repeatability in the p-value. This phenomenon is quite disconcerting because we don't often run 1000 replicate experiments, we almost always perform just one. This has to be seen as a warning to take p-values with a pinch of salt. The convention of comparing p-values to an arbitrary threshold like 5 per cent is a convenient rule of thumb—but that convenience is bought at a cost. That cost can be seen in lots of ways, but one of them is the winner's curse described here. Notice in our IQ example earlier in the chapter that the mean effect size across all our 1000 replicates is very close to the true value, the winners curse only arises when we consider only the subset of situations where p-values fell below 0.05 as suggestive that there is an effect whose size is worth evaluating. In general, we should strive to improve the power of our experiments (more on this in the next chapter), but when we have carried out an experiment we should interpret all of our explorations of the data, including the p-value, in the light of our expectation of the strength of our power. We should generally consider our estimate of effect size to be just as valid regardless of whether the p-value was above or below 0.05. In Chapter 8, we will explore the idea of optimizing experimental power not for detecting effects when they are present, but for estimating the true effect size (which might be zero) within a specified precision. For now, we would encourage you to report your estimate of effect size, not just when the p-value suggests a real effect but also when the p-value does not. Further, we would urge you to remember that if your experiment was worth doing, then it was worth doing no matter the outcome it yields, and to write up your results no matter whether they appear to suggest an effect or not. Remember our discussion earlier that failure to reject the null hypothesis should not be seen as a disappointment. Further, someone later may want to perform a meta-analysis on the range of effect sizes that have been published on studies of a particular effect—if collectively we are more likely to calculate and publish our effect sizes when the p-value suggests a real effect, then we will be biasing the data that the later meta-analysis can use. Both for your own sake and the sake of the wider scientific endeavour, remember to be just as interested in the apparent size of the effect seen in your experiment regardless of the p-value that you calculate.

 Key point

In low-powered experiments, statistically significant results are associated with over-inflated effect sizes. Take the same interest in your observed effect size regardless of the associated p-value.

2.5 Should you ever knowingly carry out a low-powered study?

To round off this chapter, we ponder one last question: should you ever carry on with a planned study, if you conclude from your power analysis that you simply cannot practically achieve high power? For example, you might be studying a potential therapeutic for a very rare disease, and this rarity constrains your sample size to a value below that required to give a power near 80 per cent for an otherwise well-designed experiment. Is it still worth doing the experiment rather than doing nothing? The answer to this is 'maybe'. If properly conducted, the experiment can add to our body of knowledge in some way—for example, it might provide useful pilot work that is part of justifying and guiding future work on this potential therapeutic, or it might not be definitive in its own right but might be useful material for a subsequent meta-analysis. However, there are still going to be costs to performing the experiment, and you are going to have to justify the value of the experiment in the light of those costs to a regulatory authority (and your own conscience), and you have to assume that these authorities will be familiar with all the arguments about problems of low power covered in this chapter. We would say that it can be possible to justify carrying out a low-powered study, but part of that justification has to be a realization that the study will be low-powered and the outcome will be carefully interpreted in the light of that.

 Key point

Sometimes it may be impossible to carry out a high-powered experiment—and you have to weigh up where you feel there would be sufficient value in carrying out a low-powered one versus conducting no study at all.

 Summary points

- One drawback to a low-powered study is that it has a high probability of failing to detect (interesting) effects.
- Another drawback is that your ability to make useful inferences from failure to reject the null hypothesis is impaired.
- If an experiment is low-powered, then a p-value suggesting rejection of the null hypothesis is more likely to be a Type I error than if the experiment is high-powered.
- In low-powered experiments, statistically significant results are associated with inflated effect sizes.
- Show the same interest in your observed effect size regardless of the associated p-value.

 Further reading

(There have been a lot of papers in the last 15 years on the dangers of low-powered studies; we have picked out a selection of those we found particularly interesting.)

Button, K.S., Ioannidis, J.P., Mokrysz, C., Nosek, B.A., Flint, J., Robinson, E.S. and Munafò, M.R., 2013. Confidence and precision increase with high statistical power. *Nature Reviews Neuroscience*, 14(8), p.585.

Button, K.S., Ioannidis, J.P., Mokrysz, C., Nosek, B.A., Flint, J., Robinson, E.S., and Munafò, M.R., 2013. Power failure: why small sample size undermines the reliability of neuroscience. *Nature Reviews Neuroscience*, 14(5), p.365.

Halsey, L.G., Curran-Everett, D., Vowler, S.L., and Drummond, G.B., 2015. The fickle P value generates irreproducible results. *Nature Methods*, 12(3), p.179.

Ioannidis, J.P., 2005. Why most published research findings are false. *PLoS Medicine*, 2(8), p.e124.

 ## Discussion questions

2.1 Why did we suggest that screening the 20 compounds for their antibiotic effects in 20 separate experiments might not be the most efficient way to proceed?

2.2 If you simply did not have the time and resources to screen all 20 seed compounds in our example with high power, then what could you do?

2.3 For your particular field of interest, discuss what fraction of tested null hypotheses are likely to be false; discuss how this value might be different in other fields that you are familiar with.

2.4 If you did want to look for an effect of handedness on IQ, what sort of sample size feels reasonable to you?

3 IMPROVING THE POWER OF AN EXPERIMENT

Learning objectives

By the end of the chapter, you should be able to explain clearly:

- Why inherent variation is a challenge to achieving good statistical power, and design decisions you can make to reduce or control the effects of inherent variation.
- How you can often increase power by changing the experimental design to increase the strength of the effect that you want to detect.
- Why increasing sample size increases power.
- Why ever higher power will be ever more costly to achieve, so you should strive for high power but not ever-higher power.

The take-home message from Chapters 1 and (especially) 2 has to be that improving the power of your experiment seems like a good idea. However, changes to your experiment that increase its power may incur other costs. These costs could be felt financially, in terms of increased effort required, and/or could involve ethical issues. Thus, in Chapter 4 we will talk about how you can estimate a value for the power of an experiment, and in Chapter 5 how you can evaluate quantitatively how different decisions about experimental design quantitatively impact on the power of your study. If you can quantify the power benefits and also quantify the costs of different options for an experiment then you have a rational basis for choosing the alternative with the most attractive trade-off between power and costs. This understanding should hopefully provide you with the toolkit you need to better design your own experiments, and to justify your design decisions to others. However, to lay the foundations for Chapter 4, here we want to consider the factors that influence power, and to discuss in conceptual rather than quantitative terms how you can make changes that should increase the power of your proposed study.

3.1   The challenge posed by inherent variation

Conceptually, we find the best way to think about improving statistical power is to appreciate the challenge that we are up against in testing any hypothesis—and that challenge is inherent variation.

If we are interested in how one characteristic of our experimental subjects (variable *A*) is affected by two other characteristics (variables *B* and *C*), then *A* is often called the response variable or dependent variable and *B* and *C* are equivalently called independent variables, independent factors, or sometimes just factors. It is likely that *A* will be affected by more than just the two factors *B* and *C* that we are interested in. Any variation in the response variable between subjects in our sample (between-individual variation) that cannot be attributed to the independent factors is equivalently called inherent variation, random variation, background variation, extraneous variation, within-treatment variation, or noise. (As an aside, we should clear up another terminological issue. In general, your scientific study will involve you taking measurements on a sample of 'things': you might be measuring lengths of strands of DNA, diameters of cells, stress hormone levels in samples of mouse blood, feeding rates in individual fish, number of breeding pairs in penguin colonies, or fractions of the human population that is under 14 years old in entire countries. So the 'things' that make up your sample can be very diverse. In this book, where we need a name for the 'things' in a sample, we will generally call them subjects, but you might also encounter them called experimental units, individuals, replicates, or participants.)

Let's use an example to see how this links to statistical power. We return to the example used at the end of the previous chapter, where we are comparing samples of IQ scores of left-handed and right-handed individuals to see if there is a difference in mean score related to handedness. The problem here is that an individual's score on an IQ test will be related to a whole bunch of things other than handedness. For example, how motivated they were to do well on the test, how well they were feeling when they sat the test, or whether the language used in the test was the language the individual routinely uses most. You could think of lots of other possible factors. All the variation due to factors we are not interested in is bundled together and called inherent variation, and that variation makes it harder to see whether there is any effect of handedness. At its simplest, if you selected a left-handed individual and a right-handed individual and found that their scores were 117 and 102 respectively—then you would not conclude that left-handed people in general have higher IQs—because these two individuals will differ from each other in lots of other ways that might reasonably be expected to potentially affect IQ. If we look at the simulations in Fig. 3.1, we assume that on average left-handers do have a slightly higher IQ than right handers (104 compared to 96), if inherent variation is such that the two distributions have a standard deviation of 15 then our power based on samples of 25 in each group is 0.48; but if we can find a way to reduce this so the standard deviation reduces to 10 then power goes up to 0.83.

Finding ways to cut through the fog of uncertainty associated with this inherent variation is the key to increasing power—and will be a running theme throughout this chapter. We will structure the chapter in terms of questions you could ask yourself in order to explore whether you can usefully improve your statistical power in any investigation you are carrying out. They will not all be useful for every investigation, but it shouldn't take you long to think them through—and the benefits might be considerable.

Fig. 3.1 (a) The distributions of 1000 left-handed individuals' IQs (mean = 104, SD = 15) and 1000 right-handed individuals' IQs (mean = 96, SD = 15). (b) The distribution of p-values resulting from Student's t-tests on 25 left-handers' and 25 right-handers' IQs, from the distributions to the left, in 1000 studies (power = 0.48). (c) As above panel, only with standard deviation reduced to 10. (d) As above panel, only using the data simulated in panel (c) (power = 0.83).

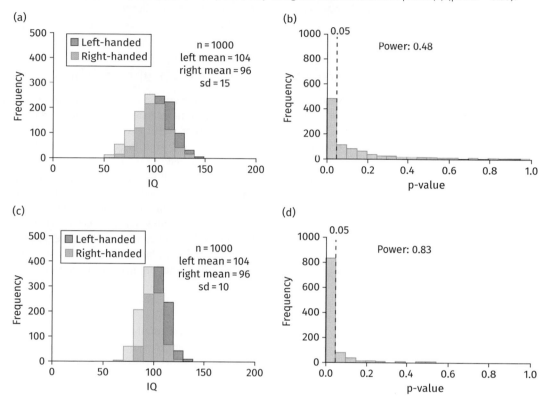

The questions you should ask are: can I usefully boost my statistical power by:

- measuring different variables?
- measuring more precisely?
- using subsampling and repeated measurement to reduce inherent variation?
- selecting experimental material so as to reduce inherent variation?
- changing the design of my experiment or the statistical analysis?
- increasing the sample size?

In the next few sections, we will look at these in more detail. It might seem odd to you that we leave increasing the size of your experiment to last, since this might well be the first solution to low statistical power that pops into your head. However, increasing the size of your experiment is definitely going to be costly, so we think it will often make sense to look at some potentially cheaper and easier solutions first. Even if you do have to increase the size of your experiment ultimately to get the power you want, you might not need to increase it by as much if you also utilize some other approaches to boosting power.

Key point

Inherent variation is a challenge to obtaining good statistical power. Our design should look to minimize and/or combat this challenge.

3.2   Are you measuring the right variables?

In our investigation into IQ we asked everyone about their handedness after they have all completed the same IQ test. It's hard to see how we could measure things differently to increase power, but it is easy to see how measuring differently to save effort could decrease power. We could save the time and effort of making people sit our standardized test if we ask them: have you ever taken an IQ test, in which case could you tell us the score you achieved? Sure, this saves effort in administering tests—but it adds noise to the data—because people might mis-remember their score, people might deliberately exaggerate their recollected score, and because those scores will have been derived from a variety of different tests taken under different conditions. We save on effort by not administering our own IQ test to subjects, but at a cost to statistical power. Our power goes down because inherent variation increases if we collect IQ data this way.

You might argue that this loss in power can be compensated for by obtaining a much bigger sample size, because people will be more willing to volunteer to answer two questions taking less than a minute than to submit to a 30-minute test; and because we can process people more quickly. We don't think we would find this a compelling argument, because the switch in measurement might increase potential for bias as well as imprecision. Let's explain what we mean here. If you think it is conceivable that IQ is linked to handedness then it doesn't seem too much of a stretch to imagine that aspects of personality such as honesty might be linked to handedness as well. Thus, if your study finds that left-handed people on average self-report higher IQs then this might be because they actually have higher IQs, or they are more likely to lie and inflate their self-reported score, or some combination of these factors. So here if we attempted to enhance power by measuring more people by allowing them to self-report IQ rather than take a standardized test, then we end up measuring something less useful: self-reported IQ rather than IQ measured on a specific standardized test. This might backfire on us. Firstly, we are adding inherent noise and (potentially more seriously) we are also increasing risk of bias, making any results hard to interpret.

We should be clear about exactly what we mean by bias and imprecision. Imprecision adds errors to your measurements but in a way that is uncorrelated from one measurement to the next. That is, if one measurement is underestimated then the next measurement might be either under- or overestimated. Bias or inaccuracy also add errors, but in a correlated way. That is, if one measurement is under-estimated by a certain amount, then the next measurement is likely to be similarly affected. In our example above, asking people to remember what their IQ was will add another source of variation to our IQ scores for each individual due to variation in individuals' recall abilities and honesty. If recall ability and honesty are unrelated to handedness, then this adds imprecision to our data, but if they are linked to handedness then this adds bias. Imprecision adds noise that makes it harder to detect any effect that we are interested in, so harms our statistical power. Bias is potentially more serious. In our example, if we thought that left-handers were on average more dishonest, and recalled on average higher IQ scores, then it is difficult to know how much of their higher IQ scores is real and how much is simply dishonest exaggeration. If you would like more discussion on bias and imprecision, then we provide some supplementary material on the website associated with this book.

Case study 3.2.1

Measurement and power

Let's look at another example. Imagine we decide to test the hypothesis that bats of a given species from higher latitude populations are on average larger-bodied. Your plan might be to obtain the necessary permissions to net, measure and release individuals at the entrance to the caves where they roost. The most obvious way to quantify size of individuals might be to weigh them. But you can imagine that an individual's weight might be quite labile—and might vary according to how recently and substantially the individual has fed, whether it has defaecated recently, its current state of heath, and reproductive state for females.

You might then wonder if measuring the wingspan of an individual would be a less labile measure of its size, and so characterizing size by wingspan rather than mass might reduce inherent variation (due to factors such as recent feeding history and health) and thus boost statistical power. This might be true or it might not. Opening up the wings of a bat to their largest extent and measuring span against a ruler is probably a more tricky procedure to master than placing a bat in a bag and suspending that bag from a spring balance to record its mass. You would want to be confident that your procedure for measuring wingspan was highly repeatable. That is, you would want to ensure that your measurement procedure did not add variation to your measurements because of variation in how you go about the procedure of opening the bat's wings to their maximum extent. A measurement has *high repeatability* if you get the same score each time that you measure the same thing, and has *low repeatability* if your scores tend to be inconsistent. Notice that high repeatability suggests low imprecision but tells you nothing about bias. You can be consistent, but consistently wrong. Again we have more in the ESM on the associated website if you feel you need to solidify your thinking on this concept.

Returning to the bats, it would be especially important to check repeatability if you planned for different people to collect the data, because they might vary in their measurement technique. However, providing you could convince yourself and others that wingspan measurements were highly repeatable, then switching from characterizing size using mass to using wingspan might buy some improvement in power—but there is no such thing as a free lunch and we have some caveats.

Firstly, you should consider the ethics of your choice of measurement variable. It may be that measuring wingspan rather than mass means that you have to hold each individual bat captive for longer. However, if you are improving power by measuring wingspan this might mean you need to catch fewer bats overall to maintain the same power as if you measured mass. This interaction of different factors influencing power is a reason why you might want to quantify improvements in power (hence the next couple of chapters in our book) before making this ethical decision.

Secondly, if we measure size using wingspan then this tells us something about structural size but there are other aspects of size that we might not pick up using this measurement. For example, if higher latitude individuals do not grow any larger skeletally but tend to carry higher fat reserves, then this is an aspect of 'size' that our measurement would not pick up. But at the same time, characterizing size by mass would not differentiate between animals being structurally larger or carrying higher fat reserves, or a combination of the two. This is a general caution about using proxy measurements, and thinking through whether they actually illuminate what you are interested in. In our case, this should cause us to ponder which particular aspects of size we are actually interested in. In the context of statistical power, we need to be careful that if we switch our proxy measurement to gain more power then this does not adversely affect our ability to answer the biological question that we are actually interested in.

Key point

What you measure and how you measure it can influence inherent variation and thus statistical power.

3.3   Can you measure variables more precisely?

It could be that some of the variation in your measurements that you are dealing with isn't actually inherently associated with the experimental material but to imprecision added by your method of measuring that material. Let's imagine in the last case study that we decided for ethical reasons to weigh bats rather than measure their wingspan as a measure of 'size'. We expect adult bats of this species to generally be 15–30 g in mass. We have in our possession spring balances that will record masses up to 50 g with precision to the nearest gramme. However, we could invest some money and buy more precise balances that record measurements to the nearest 0.1 g. This increased precision would increase our power because it is reducing uncertainty due to measurement error. Say we have two bats, bat *A* actually weighs 18.102 g and bat *B* actually weighs 17.775 g —so bat *A* is a little heavier. With our original balance we would record the weights of both bats as 18 g, so we lose in the information that *A* is heavier. But if we had bought the higher-precision balances then we record their weights as 18.1 g and 17.8 g and we recover the information that *A* is heavier. It is this unmasking of information through more precise measurement that leads to higher power.

Again, the decision to invest in the higher-precision measuring equipment will be influenced by a number of factors—most obviously in this case we will be weighing up how much of the inherent variation is likely due to measurement imprecision (and so how much our power gain will be if we invest in the better equipment) versus the financial cost of the new equipment. Other issues can come into play too—it might be that the more precise scales are less practical because they are more fragile and vulnerable to damage under field conditions and/or need unfeasibly regular recalibration to avoid bias. In this regard, it is as well to bear in mind that when you use scientific equipment in practice you definitely don't always achieve the measurement precision that manufacturers boast about in their promotional material.

We should also remind you that improving precision isn't always about buying more expensive equipment, it is often about improving your protocols for using the existing equipment—so you increase the standardization (and so repeatability) of your measurements. In the case of the bat experiment we might provide written instructions for all measurers to check the performance of their scales at prescribed intervals using objects of known mass, and instruct them on how to clean the bag prior to using it to hold each bat.

One way to improve power is to replace a qualitative measurement with a quantitative one. Imagine we wanted to explore whether being overweight was linked to allergies and we asked people attending an allergies clinic if they would be willing to take part in the study and, if so, whether they considered themselves overweight or not. We could then compare the fraction of self-reported overweight individuals with that of the general population to test our hypothesis. You can see that there is an issue with bias here and people's self-reporting may not be reliable, but there is an issue of precision too—with everyone who takes part simply being assigned to one of two categories: 'overweight' or 'not'. Instead we might ask each potential volunteer if they are willing to have their height and weight measured, from which we can calculate Body Mass Index. This is a little bit more work, but we come away with a much more fine-scaled measure of the person's body; this increase in quality of information should lead to higher statistical power. (We provide a bit more discussion of different types of measurements in the EMS.)

Key point

Increasing the precision of your measurements can be another means to increase power.

3.4   Repeated measurement and subsampling to reduce inherent variation

Sometimes you know that your measurement will be adding noise but there is not really a convenient way to reduce that noise using the methods discussed in the last section. In this situation repeated measurement and taking the average might be a good idea. Let's imagine that we want to compare counts of seals at ten coastal colonies near areas where fishing boats have been excluded versus ten colonies where fishermen are active nearby. For each site we have eight aerial photographs of the colony, and we want to count the number of seals that we see on each photo. There is a real chance of imprecision here—with seals being overlooked or shadows or rocks being mistaken for seals. It might be rational for us to call in five student volunteers and train them up in how to spot seals in photos then ask each independently to report back counts of all the 160 photographs. Following these repeated measurements, for each photo we can then use the average of the five scores. If mistakes in scoring add imprecision then using the mean of several repeated measurements of the number of seals in each particular photo should reduce this imprecision, and so enhance power. Of course, the cost here is in the extra person-power required. We should also be careful to avoid a situation where drawing in extra personnel to tackle this issue causes us to use a scorer who adds bias to the measurement procedure (e.g. if one of our five students consistently misses a lot of seals). However, we can test for this—since we would expect that on average the five individuals have similar distributions of scores.

Subsampling is a similar concept to repeated measurement but here, rather than repeatedly measuring the same experimental material, we measure a number of randomly selected parts of each individual that we want to sample and take the mean of these as a representation of that experimental unit. Imagine we wanted to compare the ectoparasite levels of chicks of a species of bird nesting in nest boxes versus natural cavities. We would expect nestlings in the same brood to be similar in ectoparasite load but not identical, since they will share the same environment and be in close proximity to each other—but they will not be identical in their genetics, in their health, or in growth rate, all of which might influence their susceptibility to parasites. The experimental unit in this study is a particular nest. We might simply randomly select one chick from each nest and count the number of ectoparasites on it as a representative measure for that nest. But sometimes we might select an individual with relatively strong immunity and sometimes one with weak immunity, this will add intrinsic variation and reduce our power. To combat this, we might decide to sample four random chicks from within the nest (or all of them if their number is lower) and then record the average number of ectoparasites per chick in that nest. This costs us more effort, and our sample size does not increase for this extra effort because the sample size is the number of nests. However, this extra effort might give us a boost in power if using the mean number of ectoparasites from several sampled individuals reduces intrinsic variation caused by differences between chicks that share the same nest. The techniques we will introduce in the next two chapters would allow you to explore quantitatively the power consequences of different designs in terms of how many chicks you assay per nest. This procedure is called subsampling, because you do not measure all the experimental material in each experimental unit (i.e. all the chicks in a nest), but only a randomly selected subset of them.

We might have considered subsampling in our seal colony investigation. Rather than attempting to count all of the seals in one photograph of a colony we might have reduced our effort and instead placed a regular square mesh across the photograph and only counted seals in a random sample of squares in our photograph—multiplying up to get an estimate of the total population size. By doing this we potentially reduce power because we are introducing intrinsic variation when we replace the total counts with estimates produced by subsampling. However, it is possible that we end up with more precise estimates this way because we study the small sections much more carefully than the whole photo and make fewer counting mistakes. So, subsampling could actually lead to an improvement in power if it increases measurement precision—a pilot study combined with the techniques for estimating power outlined in our later chapters would allow you to evaluate the best approach for your situation.

 Key point

Repeated measures and subsampling can increase statistical power.

3.5 Can you select experimental material so as to reduce inherent variation?

One way to reduce intrinsic variation is to select our sample material so as to standardize it for variables that we fear might be adding intrinsic variation. For example, we might think that occupation, extent of education and age might all strongly influence IQ, so in our investigation of handedness we might decide to only include people in our study aged 18–22 who are studying at university. This should increase our power because we reduce the intrinsic variation between individuals caused by these variables (age, occupation, and extent of education). But what are the costs to this approach? Well, one cost might be that we have to work a little harder at recruitment now we are being more fussy about who we are taking into our study. A more serious implication is that we have subtly changed the hypothesis that we are testing and this might impact on how comfortable we are generalizing the outcome of our study. We originally asked the question whether handedness affected IQ; we have now narrowed our enquiry to 'does handedness affect IQ in individuals aged 18–22 in higher education?' Now it seems likely that if we do find an effect in our narrower study, that this effect would hold (at least qualitatively) in people more broadly. However, we do not have the data to support this wider assertion—if we wanted to make this suggestion, we would be doing so by appealing to biological reasoning and intuition rather than hard evidence.

This trade-off occurs more widely. Generally, the reason we conduct some experiments in the laboratory is not because we are particularly interested in how animals respond under laboratory conditions but because we can control conditions more carefully in the laboratory. One advantage of that (compared to a similar experiment performed in more natural conditions) is that we have fewer factors contributing intrinsic variation and thus we have higher power; but the drawback is less surety in how our results generalize to natural conditions (which are often the conditions that we are especially interested in). To introduce you to some jargon, people describe a study as having low external validity if it is based on a sample drawn from a narrower population than you are really interested in (e.g. a sample of students when you are interested in people more generally).

It is for analogous reasons that we sometimes decide to conduct experiments *in vitro*, we sometimes use laboratory-bred populations of test subjects, and we sometimes use inbred strains and even clones to reduce or remove intrinsic variation due to genetic factors. In each case, there are likely benefits in terms of increased power but costs in ease of generalization.

 Key point

You can select your experimental subjects to be more standardized to reduce inherent variation and thus increase power, but at the cost of reduced generalizability.

3.6   Can you strengthen the effect that you are interested in?

Up to this point, we have been talking about ways to reduce variation due to other factors so we can see the effect that we are interested in more clearly. We can make the effect easier to detect by designing the experiment so that this effect (if it exists) is likely to be stronger. As an example, Fig. 3.2 is analogous to Fig. 3.1 but now we see that increasing the difference in the mean IQ between the two groups increases power.

Consider again our interest in seeing if bats of a given species increase in size with increasing latitude. We can maximize our chance of detecting any such effects by making the populations that we are comparing as different in latitude as possible. Say we have decided that we will sample 12 different populations—we strengthen our power if we compare a group of six populations from really low latitudes with six populations from the really high

Fig. 3.2 (a) The distributions of 1000 left-handed individuals' IQs (mean = 104, SD = 15) and 1000 right-handed individuals' IQs (mean = 96, SD = 15). (b) The distribution of p-values resulting from Student's t-tests on 25 left-handers' and 25 right-handers' IQs, from the distributions to the left, in 1000 studies (power = 0.51). (c) As above panel, only with the effect size increased: left-handed-individuals' mean IQs = 108, right-handed individuals' mean IQs = 93, SD = 15. (d) As above panel, only using the data simulated in panel (c) (power = 0.9).

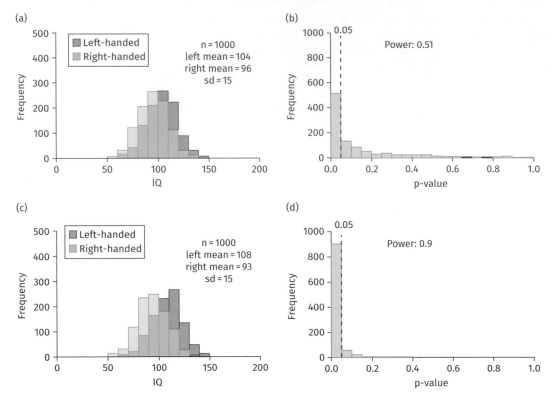

latitude end of the species' range (rather than spacing the 12 populations out across all latitudes in the range). However, we should warn you that there is another trade-off here. Really, we have changed the hypothesis under investigation a little bit and we are now asking: are bats from the extreme low-latitude part of the range different in size from those at the extreme high-latitude end? We end up learning nothing about the mid-latitude individuals. Another way to look at this is that by increasing our potential to detect any sort of effect if it exists, we lose our ability to learn about the nature of the effect. Let's say we do find that the high-latitude individuals are bigger than the low-latitude individuals; we don't know if the change occurs linearly as we move to higher and higher latitudes, or if there is some sort of non-linear trend. Indeed, for all we know the mid-latitude bats could be even bigger than the high-latitude. So you would have to weigh up what is more important to you, maximizing your chance of definitely detecting an effect if it exists, or characterizing the nature of any effect you find. Of course, compromises could be adopted, and you could have perhaps a group of four populations at each extreme and another four scattered across mid-latitudes—the methods we will introduce you to in the next two chapters would let you explore power trade-offs of these different designs quantitatively.

Strengthening the effect of interest is easier in a manipulative study. Imagine we wanted to explore the beneficial impact of dusting crop seeds with an anti-fungal agent on eventual yield. We would likely compare fields planted with undusted seeds (or seeds dusted with an inert material) with fields sown with seeds that have been dusted with the anti-fungal agent. The stronger the concentration of active ingredient that we use in the powder that we mix with seeds, the stronger we would expect the effect to be. Hence, your power to detect an effect will increase with higher concentrations. Thus, there would be a temptation for you to use a very high concentration in your study. The compromise here is to be careful that in the quest to boost power you do not end up adopting an unrealistic manipulation. In the example here, evidence of an increase in crop yield would be of less interest if it was achieved using concentrations of active ingredient that were too high to be economically viable when employed commercially, or that posed an environmental hazard, for example.

 Key point

We can often increase power by changing the experimental design to increase the strength of the effect that we want to detect. but remember that this may change the hypothesis under test in subtle, but potentially important, ways.

3.6.1 Asking yourself what the minimum effect size is that you are interested in

In the previous section we discussed how you could change your experiment so as to increase the effect size (if it exists) as a means of increasing power. But for any study you should ask yourself what is the smallest effect size you are interested in being able to detect. This may seem odd to you, and you would be tempted to answer that you would be interested in knowing about any non-zero effect, no matter how small. But is this really true? If we told you that high latitude bats where on average 0.001 per cent heavier than low-latitude ones—would you really find this difference interesting? We think probably not. As we will discuss fully in the next chapter, you need to specify this minimum effect size that you are interested in to calculate power. The higher this minimum the higher your power; intuitively you can see that it is easier to detect big effects than small ones. So you do have to think about this minimum value: the lower you go, the more it will hurt your statistical power. In upcoming chapters, we will offer you the tools to describe this trade-off for your particular study quantitatively. When you have that trade-off curve, that can help you decide on how big an effect has to be for you to want to have good power to detect it.

 Key point

To calculate power, we have to specify the minimum size of effect we are interested in being able to detect—the higher the minimum value we can live with, the greater our power.

Case study 3.6.2

Using standardized effect sizes

In this book, when we talk about effect sizes, we are talking about *unstandardized* effect sizes where the effect size generally has units attached to it. When you read other works, you will sometimes see standardized effect sizes, where the effect size is scaled to some measure of variability such that it becomes a dimensionless number, free of units.

As an example, imagine that we surveyed cigarette use in co-habiting mixed-sex couples. We might report the observed effect of sex on cigarette use as follows: '*on average the male smoked 2.3 cigarettes a day more than the female.*' This is the unstandardized effect size, it is measured in numbers of cigarettes. However, another way we might have expressed this effect size is to use something called *Cohen's d*. This is a standardized measure of effect size that is often used when two groups are being compared, and it is obtained by dividing the difference in the means between the two groups by an estimate of the pooled standard deviation. You can think of it as the difference per unit of variation. So you might see this reported as follows: '*in our survey of cigarette use we found a small effect of sex, with males using more (Cohen's d = 0.22).*'

Of course, there is nothing to stop you quoting the results of an experiment in both unstandardized and standardized forms. Each have their strengths. We have used unstandardized effect sizes here because we think that they are more easily interpreted biologically providing the units of measurement are relatively intuitive. In our case above, it is easy for us to grasp what two cigarettes extra smoked per day might be like and consequently how biologically important such a difference might be. So, from our perspective, when considering the effect sizes to use in power calculations, we find it easier to discuss unstandardized effect sizes. For example, if we are looking at investigating the effect of using nicotine patches in reducing cigarette consumption, we might argue that the minimum effect size that we would be interested in detecting was one cigarette per day less in the treatment group compared to controls. In evaluating our power analysis, we think it is easy for you to grasp this unstandardized effect size. Whereas if we said that our minimum standardized effect size of interest is 0.1 then we think this is harder to envisage.

The main attraction of standardized effect sizes is that they allow comparison across studies much more easily. Imagine that our study in nicotine patches reported cigarettes smoked per day, whereas another study reported cigarettes bought per week, another estimated tar consumed (to take account of different brands of cigarette), and yet another asked participants to record their average level of craving for a cigarette each day on a 0–5 scale. Although the studies might be similar in design—comparing a group of individuals using nicotine patches with another group using placebos—the different approaches to measuring effectiveness of the patches makes comparison between the groups difficult. This difficulty is greatly reduced if a conventional standardized measure (most obviously Cohen's d in this case) is reported for each study. If we reported that d-values for the four studies are 0.23, 0.19, 0.24, 0.56 then we can see at a glance that the last study seems to suggest a stronger effect of the patches than the other three studies. Hence, standardized measures of effect size are very useful when comparing the outcome of a diversity of similar studies, and thus are pivotal to meta-analyses.

Historically, standardized effect sizes have also allowed the production of tables of power values, analogous to the statistical tables found at the back of older statistics textbooks. If you work with unstandardized effect sizes then you would need a different table for every type of measurement, but if you convert your effect size to a standard scale you can use the same power table (or more recently, online power calculator) no matter what the biology of your study. Of course, if you are writing bespoke simulations as we have in this book, this advantage no longer applies.

There is another way that standardized effect sizes can be useful. For commonly used standardized effect sizes, conventions have built up over time as to the interpretation of the value of a standardized effect size. You might have noticed this above where we referred to a Cohen's d value of 0.22 as a small effect size. There is a convention for Cohen's d than around 0.2 might be regarded as a 'small' effect, 0.5 'medium' and 0.8 'large'. So, you might use this convention as part of your consideration of what sort of effect sizes are of interest in your study. But our view is that this will only be occasionally useful to you as a way of starting off your deliberations, because the size of effects that might be considered 'small' in the context of one area of scientific enquiry might be considered 'large' for another—and this general convention does not take that into account. Consider whether a 10 per cent increase in your salary would change your life more than a 10 per cent increase in your body-weight or a 20 per cent increase in the number of novels you read. However, if you can find the standardized effect sizes reported in previous similar studies to yours, then this might be a useful basis for selecting the effect sizes that you expect in your study, even though your study might be taking quite a different measurement approach to the previous studies. So, standardized effect sizes can be useful in power analysis if you are using previous similar studies as a basis for justifying the strength of effect that you are using in your power analyses. Cohen's d is by no means the only commonly used standardized effect size. Most books of meta-analyses will provide an effective overview of alternatives. We particularly recommend Grissom and Kim 2005 and Cummings 2013.

 Key point

Standardized effect sizes aid comparison between studies; but unstandardized ones can be easier to interpret biologically.

3.7   Can you change the design of your experiment to boost power?

So far in this book we have concentrated on the simplest of experiments, where we are comparing between groups of individuals. More complex designs are often adopted in order to manage inherent variation, and switching to a more complex design can sometimes offer an increase in power. In the anti-fungal seed dusting experiment discussed earlier in the chapter there will be a lot of field-to-field variation—some fields will have better water-draining qualities, or an aspect that offers more sunlight, or soil high in essential nutrients. In the simplest experiment we would identify say 14 fields and randomly allocate seven of them to receive the active-ingredient dusting and seven to experience the control treatment. Essentially, we are counting on our randomization to mean that all of the other aspects that might affect yield are approximately equal on average across our two groups (even if they will differ from field to field). This means there will be a lot of inherent variation that will cloud our ability to detect a difference between the two groups on average. We could change the design in a number of ways that might improve power by dealing with inherent variation in more sophisticated ways.

For example, we could switch to a within-subject design. Now we will have to run our experiment over two growing seasons. Each field will experience both the active and the control treatment in different seasons (with half of them experiencing the active one first, and half the active one second). We could then look at the difference within each individual field in crop yield with the active ingredient versus the control. Here, because we are comparing within a field, the differences between fields don't have the same masking effect.

It is almost certain that there will be historical records of yield from each of the fields. This can be used to boost our power even if we only have one growing season to run our experiment over. For example, we could use field 'previous performance' as a blocking factor. We could first split the fields into poor-, medium-, and high-quality subsets, randomize within each of these three blocks, and compare within these blocks. Since we are always comparing 'similar quality' fields under the different treatments, then the other factors that influence yield should be less of a problem. Alternatively, we could carry out our original experiment but make our statistical analysis a bit more refined, by using our measure of field 'previous performance' as a covariate in our analysis.

As ever, the decision to go for a more complex design or analysis to boost power is not a given. There are costs to added complexity. You can see that we have to be more patient and wait two seasons instead of one to get our results if we adopt the within-subject design. You can imagine that it is easier to make mistakes, and harder to explain your results to others, if you adopt a more complex statistical analysis. Blocking on a factor that does not actually contribute a substantial fraction of the inherent variation can actually cost you statistical power. However, our general point stands, more complex designs can sometimes be attractive as a way of boosting statistical power.

 Key point

Sometimes adopting a more complex experimental design can help account for inherent variation and so increase power.

3.8   Would you be willing to accept a higher rate of Type I error?

Remember from Chapter 1 that the convention that a p-value below 0.05 implies a significant result (evidence to support rejection of the null hypothesis of no effect), is just a convention. You could adopt a higher value for this criterion, say 0.1 rather than 0.05. The upshot of switching to 0.1 is that we will double our Type I error rate. That is, if there truly is no effect to discover, our sampling and statistical testing on that sample will erroneously suggest that there is an effect 10 per cent of the time (rather than 5 per cent). However, the other consequence of this change is that we will also increase our power to detect an effect if there really is an effect to be detected. This might be an attractive trade-off to you if you are in a situation where false positives do not bother you overly but false negatives (missing interesting effects) bothers you a lot. An example for this might be a situation where we are screening lots of biological compounds to see if any of them have antibiotic properties that might be the basis of a drug treatment. Let's assume that we have a relatively quick and cheap assay; then we can screen lots of compounds. Our expectation is that most compounds will not be antibiotics, but that if we do find one then that might be a very useful discovery for us. This is exactly the sort of situation where we might relax our Type I error rate to boost power.

Imagine that we test 1000 compounds, of which only two are actually antibiotic. It might be that if we have a 0.05 threshold then our initial pass flags up 48 compounds as being of interest of which one is one of the truly antibiotic. With a 0.1 threshold we flag up 113 as being of interest, including both the truly antibiotic compounds. In each case, after the initial screen, we subject all the potentially interesting compounds to a more expensive and time-consuming (but more reliable) investigation. In the case with the 0.05 threshold we only need to put in half the effort into the second screen and we discover one of the antibiotic compounds; with the 0.1 criteria we do have to invest more in our second phase but we end up identifying both of the potential antibiotics. Given that antibiotic compounds are rare, we might well be more comfortable with the second scenario.

 Key point

Increasing the Type I error rate that you set will increase your power.

3.9   Can you increase sample size?

As we discussed in Chapter 1, the challenge of statistical testing is that we want to know something about some population, but it is impractical to measure the entire population, so we only measure a sample of that population and then make inferences about the overall population on the basis of what we see in the sample. The smaller the sample the less of a precise reflection of the population it is, and the more random effects of sampling (coupled with inherent variation) can cause the sample to deviate in its properties from the population. Thus, intuitively the larger our sample size the more our sample will reliably reflect the population. Thus, with a larger sample, if there really is an effect of the type that we are interested in detecting in the population then we will be more likely to pick up that effect in our investigation of the sample. To put it another way, increasing sample sizes increases statistical power (as you see in our Fig. 3.3).

It may seem perverse of us to discuss a bunch of other ways to increase power before mentioning increasing sample size, since this was probably the strategy for increasing power that would first pop into your head. But we think it might be rational for you to consider all the other approaches first before increasing sample size. Almost invariably increasing sample size is going to cost you time and financial outlay, it might require you to find additional housing and maintenance resources for your increased sample, and it is likely to have ethical implications. Ethical considerations generally produce pressure to reduce the size of experiments, providing this reduction does not seriously compromise the value of the experiment—thus, almost inevitably increasing sample size to increase power will need to be very carefully justified ethically. We are not saying that you won't ever be able to justify this increase, but part of that justification ought to be that you have also considered other means of improving power—so we want to encourage you not to have increasing sample size as your default strategy for improving power; you can be much more creative than that.

 Key point

Increasing sample size increases power.

Fig. 3.3 (a) the distributions of 1000 left-handed individuals' IQs (mean = 104, SD = 15) and 1000 right-handed individuals' IQs (mean = 96, SD = 15); (b) the distribution of p-values resulting from Student's t-tests on 25 left-handers' and 25 right-handers' IQs, from the distributions to the left, in 1000 studies (power = 0.44); (c) the distribution of p-values resulting from Student's t-tests on 35 left-handers' and 35 right-handers' IQs, from the distributions to the left, in 1000 studies (power = 0.63).

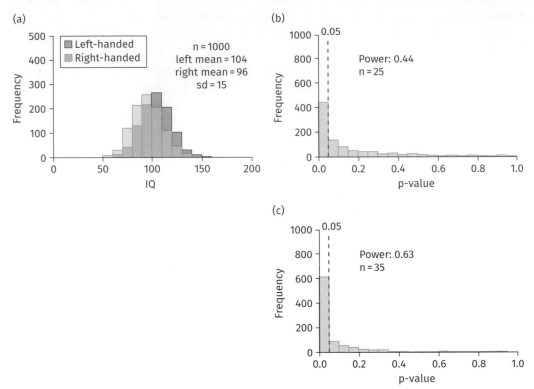

3.10   Practical and ethical reasons why you should not always seek to further increase power

You can see from the discussions so far in this chapter that there are trade-offs involved in any approach you take to increasing power. You won't be surprised to also find in the next chapter that there is an effect of diminishing returns in terms of power increase from increasing effort to boost power. As a taster for this, if we were looking to estimate the mean mass of bats of the species we discussed earlier in this chapter, then we could draw a random sample of size 100 and work out the mean of those. Now this mean will be close to, but not exactly the same as, the population value we would get if we (hypothetically) measured every single bat. The difference between the estimate from the sample and the true value tends to scale with the inverse of the square root of the sample size (this result comes from a body of statistical theory called the Central Limit Theorem—if you would like to chase it up). Now for our sample of 100, this inverse of the square root of the sample size is 0.1. If we wanted to double our precision (i.e. half this distance), then we would want the inverse of the square root to be 0.05, so the square root of the sample size to be 20: and so the sample size to be 400. Thus, to improve our precision by a factor of 2 we have had to increase our sample size by a factor of 4 (from 100 to 400). It won't surprise you that, by the same reasoning, to improve our precision by a factor of 4 we would need to increase the sample size by a factor of 16. By similar reasoning, if we want to increase power by some fixed amount, then we will need to invest more to achieve this uplift if we already have a moderately powerful design than is we are starting from a low-powered design. For this reason, we cannot practically strive for power of one. But how much is good enough?

Actually, this is an interesting question that we feel is worth careful exploration—so we return to it in the next section—but for now let's offer a rule of thumb. There is a convention that power of 80 per cent is reasonable. We need to stress that this is arbitrary (like the 5 per cent convention for Type I error rate). There may be times when you could have good reason to deviate from this 80 per cent convention, and that is what we want to take more time to explore in the next section. For now, we should introduce you to the 80 per cent convention because you will meet it time and again—but remember it is nothing but a convention. Conventions exist because they are useful much of the time, but we always need to be on our guard for situations where adopting the convention is not the best idea (but let's leave that for the next section).

💡 Key point

Ever more power will be ever more costly to achieve, so we should strive for high power but not ever higher power.

Case study 3.11

Thinking a bit more about how much power is enough

In the last section we emphasized that the 80 per cent power criterion is nothing more than a convention. This convention can be traced back to the influence of the late Jacob Cohen, who has been the dominant force in raising the profile of power analysis among practicing scientists. We should emphasize that Cohen did not argue forcefully for adoption of this convention, he simply used it for illustration, but such was the influence of his work in the development of the practice of power analysis it has become a convention.

We are familiar with the 5 per cent convention for the Type I error rate (α) that comes about by adopting a critical P value of 0.05 for rejection of the null hypothesis. You will readily find literature arguing that in some situations there may be good reason to deviate from this convention. For example, in situations involving mass testing of hypotheses (often in the context of genomic studies) it can be argued that a much lower threshold must be set in the context of carrying out a huge number of tests (Day-Williams and Zeggini 2011). In the context of smaller numbers of related tests, you will find a literature on the reduction of α in individual tests so as to control Type I error rate at some higher level than that of an individual test (Benjamini and Hochberg 1995), but the philosophical and practical challenges to implementing these can often be non-trivial (e.g. Nakagawa 2004; Glickman et al. 2014).

Conversely, we have argued earlier in the chapter that a higher α value might sometimes be rational in the context of boosting power. Also, where a study is carried out in an exploratory philosophy of generating questions rather than a confirmatory philosophy of testing specific hypotheses (see Tukey 1980) then setting a higher α level might again be rational. There has been much less published discussion of purposefully deviating from the power = 80 per cent (i.e. β = 0.2) convention.

You might find it argued (e.g. Di Stefano 2003) that adopting $\alpha = 0.05$ and $\beta = 0.2$ (i) implicitly assumes that the cost of making a Type I error (a false positive FP) is four times that of making a Type II error (a false negative FN), and (ii) that it would be more rational to first consider the ratio of these costs in your specific circumstances when setting your desired power level. There is some truth in this, but really the situation is more complex.

Statement (i) is only true if we assume that (a) the probability of there actually being an underlying effect to discover is 50 per cent (i.e. the parameter R introduced in Chapter 2 is one) and (b) successfully detecting a real affect (a true positive TP) and correctly not suggesting there is an effect (a true negative TN) are equally valuable to you. The first of these is problematic—R is unknowable but there is no reason to think it is consistently around one. The second issue is interesting. On one hand you could say that science is the search for truth, and so the two are equally valuable. However, you only have to delve into the abundant literature on publication bias away from publishing negative results to know that this ideal will not always be met. Thus, to justify moving away from the 80 per cent standard you will need to think about estimating what R is likely to be for your field of inquiry and about the relative benefits of discovering true positives and true negatives. Moreover, you will also need to make an assumption about how the scale of the benefits of these two types of successes relate to the costs of making the two types of mistakes.

This sort of investigation is perfectly possible. Burt et al. (2017) perform an interesting simulation analysis in the context of clinical trials. New medical treatments are often considered to go through three sets of human testing: phase 1 on small samples to check that they are not harmful, phase II on larger samples to evaluate short-term efficacy, and phase III on larger samples yet to evaluate long-term effects. Burt et al. argued that phase II clinical trials should have greater statistical power than is currently commonplace. Their argument is that there is a very high opportunity cost to missing genuine treatment effects, because developing a successful therapy can offer huge financial returns. These additional returns more than make up for the fact that these larger phase II trials will be more expensive and will lead to a need for increased number of (by definition larger and more expensive) phase III trials to follow-up on the increasing number of candidate therapies that will pass the more powerful phase II trials. However, it should be remembered that the costs of clinical trials are not just financial—any change of this nature would need to carefully weigh up ethical as well as financial dimensions—and consider wider societal good versus the potential costs to those involved in trials.

This leads us to another challenging aspect of calculations such as these to justify the target power level that you adopt. In terms of quantifying the costs and benefits of true and false positives and negatives, we have to acknowledge that different stakeholder groups might have different perspectives on these (see Ioannidis et al. 2013). In the example of clinical trials you can see that the personal ambitions of those carrying out the trials, financial strategists of the companies involved, patient groups, regulatory authorities, and politicians might have quite different viewpoints.

None of this is meant to suggest that you should just accept the 0.8 convention. There will be times when a different power would be more appropriate. What we are saying is that if you are going to argue for designing your experiment to have a power different from 0.8 then you are going to have to be prepared to present a compelling case because you will have to convince others for whom the 0.8 convention is entrenched. Making a compelling case is possible, given the tools offered in this book, combined with arguments about likely values for R and the costs and benefits of false and correct positives and negatives. Producing 'likely values' that others will agree to will be challenging and you will likely have to perform simulations to explore a range of possible scenarios. You may also have to accept that these 'likely values' are not absolutes but will differ according to the perspectives of different people that you have to convince. There is no getting around it, this is going to take a long time (and will involve you in psychology, ethics, economics, and politics, as well as science)—the easy thing to do is give up the fight and stick with 0.8. However, if you are involved in really high-stakes activities (e.g. you are advising the regulatory authorities overseeing clinical research) then you really cannot afford to take the easy way out.

 Key point

Deviating from the 80 per cent power convention will require some justification, but sometimes you might have to grasp that nettle.

Summary Points

- Higher statistical power is attractive but will come at a cost; good experimental design is about selecting the experiment with the best trade-off between power and costs.
- Inherent variation is a challenge to obtaining good statistical power. Your design should look to minimize and/or combat this challenge.
- The traits that you decide to measure on experimental subjects and how you measure them can influence inherent variation and thus statistical power.
- Increasing the precision of your measurements can increase power.
- Repeated measures and subsampling can increase statistical power.
- You can select your experimental subjects to be more standardized to reduce inherent variation and thus increase power; but this may hurt your ability to generalize from your study.
- We can often increase power by changing the experimental design to increase the strength of the effect that we want to detect; and/or by only being interested in detecting larger-sized effects.
- Sometimes adopting a more complex experimental design can help account for inherent variation and so boost power.
- Increasing the Type I error rate that you set will increase your power.
- Increasing sample size increases power.
- Ever more power will be ever more costly to achieve, so we should strive for high power but not ever higher power.

Further reading

Benjamini, Y. and Hochberg, Y., 1995. Controlling the false discovery rate: a practical and powerful approach to multiple testing. *Journal of the Royal Statistical Society.* Series B (Methodological), pp.289–300. [Discussion of Type I control in the context of multiple testing.]

Burt, T., Button, K.S., Thom, H.H.Z., Noveck, R.J. and Munafò, M.R., 2017. The burden of the 'false-negatives' in clinical development: Analyses of current and alternative scenarios and corrective measures. *Clinical and Translational Science*, 10(6), pp.470–479. [Interesting discussion of power in the context of different phases of clinical trials.]

Cohen, J., 1988. *Statistical Power Analysis for the Behavioural Sciences* (2nd edition) Laurence Erlbaum, NJ. [Quite simply, the most often cited text on statistical power.]

Cumming, G., 2013. *Understanding the New Statistics: Effect Sizes, Confidence Intervals, and Meta-Analysis.* Routledge. [Good introduction to standardized effect sizes.]

Day-Williams, A.G. and Zeggini, E., 2011. The effect of next-generation sequencing technology on complex trait research. *European Journal of Clinical Investigation*, 41(5), pp.561–567. [A useful discussion of the statistical consequences of mass testing of hypotheses.]

Di Stefano, J., 2003. How much power is enough? Against the development of an arbitrary convention for statistical power calculations. *Functional*

Ecology, 17(5), pp.707–709. [Discussed reasons to deviate from the 80 per cent power convention.]

Glickman, M.E., Rao, S.R. and Schultz, M.R., 2014. False discovery rate control is a recommended alternative to Bonferroni-type adjustments in health studies. *Journal of Clinical Epidemiology*, 67(8), pp.850–857. [Discussion of Type I control in the context of multiple testing.]

Grissom, R.J. and Kim, J.J., 2005. *Effect Sizes for Research: A Broad Practical Approach*. Lawrence Erlbaum Associates Publishers. [Good introduction to standardized effect sizes.]

Ioannidis, J.P., Hozo, I. and Djulbegovic, B., 2013. Optimal Type I and Type II error pairs when the available sample size is fixed. *Journal of Clinical Epidemiology*, 66(8), pp.903–910. [A thoughtful discussion of the costs of errors.]

Nakagawa, S., 2004. A farewell to Bonferroni: the problems of low statistical power and publication bias. *Behavioral Ecology*, 15(6), pp.1044–1045. [Discussion of Type I control in the context of multiple testing.]

Ruxton, G. and Colegrave, N., 2011. *Experimental Design for the Life Sciences*. Oxford University Press. [It does seem cheesy to recommend our other book. But this is the chapter of this book that has most commonality with our previous one. Many of the issues in this chapter are explored in more detail in that book.]

Tukey, J.W., 1980. We need both exploratory and confirmatory. *The American Statistician*, 34(1), pp.23–25. [A discussion on exploratory versus confirmatory philosophies behind hypothesis testing.]

Discussion questions

3.1 What factors might you expect to be strongly linked to someone's IQ?

3.2 How might you seek to improve the precision of the student volunteers in the seal count example?

3.3 In what order should the students score the 160 photos?

3.4 In the seed-dusting experiment run over two years, why would we not give all the fields one treatment in the first year, then all of them the other treatment the next year?

4 HOW TO QUANTIFY POWER BY SIMULATION

Learning objectives

By the end of the chapter, you should be able to explain clearly:

- What you need to know (or be able to estimate) about a potential study in order to estimate its power.

- How (conceptually) to take these assumptions and convert them into a good estimate of statistical power.

- How to convert that conceptual understanding to being able to achieve this practically in R (or any other programming language you are familiar with).

- How to modify your approach to compare multiple possible versions of a study in order to decide which one offers the power you need most efficiently.

Now that you understand what power is (Chapter 1), what factors influence power (Chapter 3), and the importance of ensuring your planned study has high power (Chapter 2), how do you go about estimating the power of a study yourself? Answering this is the central focus of this chapter. We will work through a specific example involving a simple experimental design. This example will allow us to deal with the key aspects involved in estimating power that you can apply to any study. You will also learn some of the basic computational skills that will form part of the toolkit you will need to perform power analyses using the method we suggest. Later in the book we will show how the same fundamental approach you will learn here can be adapted to deal with any type of study design that you might be considering. But let's get the fundamentals sorted out first.

4.1   What you need to know to estimate power, and ways to produce plausible estimates of these

In Chapter 3 we saw that the power of any study is affected by a number of factors that we can organize into three groups: how you plan to design the study; how strong the effect we are interested in detecting is likely to be (if it exists at all); and how strong inherent variation is likely to be.

Let's start with the design of the study. This includes the planned sample size, the way you will allocate subjects within your study to different treatments, and the analysis you plan to carry out at the end. It also includes a decision about the p-value threshold you will use to decide whether to reject your null hypothesis (called Type I error rate and often denoted α). By convention this is often 0.05, but this value is not set in stone. The next thing we need to think about is the actual size of the biological effect that is there to be detected. What this effect is will depend on the type of study that you are carrying out. In this chapter we are going to focus on a simple study with two treatment groups, and so the effect size will be the change in the average value of the measurement of interest caused by the treatment. In other types of study, the effect size might be the slope of the relationship between two continuous variables or the strength of their correlation. The third and final thing to think about is what the biological variation in your study system looks like.

When setting out to calculate the power of a study, you will generally know the plan that you have in mind for your study design (or at least have a list of possibilities to consider . . . if not, we can recommend a very good book on experimental design for the life sciences!). However, you will not know the effect size, because if you did, you would not need to do the study! You may not know exactly what inherent variation you are faced with either. This does not seem like a great start . . . Of the three things you need to know to estimate power, you will actually often only know one. Fortunately, there are ways to proceed, otherwise this would be a very short book. So let's think in more detail about effect size and inherent variation.

4.1.1   Effect size

The fact that you don't know the real biological effect leaves you with a problem; you cannot determine the power of your study to detect that actual effect. The solution is not to try. Instead, you are going to estimate something slightly different. You will decide on the size of the effect that you want your study to be able to detect, and then estimate your power to detect an effect of that size (if it exists). It is at this point that many people become uncomfortable . . . surely simply picking your own apparently arbitrary effect size out of thin air means that the power analysis is meaningless? But of course you are not going to pick it out of thin air, you are going to make use of your biological knowledge of the system, you will consider the broader aims of your study and then spend a serious amount of time and thought coming up with the smallest effect size that you want your study to be able to detect (or at least have high power to detect). And then you estimate your planned study's power to detect an effect of that size. In fact, one of the major additional benefits of carrying out power analysis is that it forces you to focus very clearly on exactly what you are trying to achieve with your planned study.

For example, if your study aims to investigate the effect of a new dog food supplement on the health of dogs, you might only really be interested in being able to detect benefits that are at least as big as those of the best currently available food supplement, which offers a 5 per cent increase in longevity. In that case, you might set this level as the minimum effect you want to have a high power to detect. In doing so, you accept the fact that if the supplement actually has a much smaller effect than this, you are unlikely to detect it with your study, but this does not matter to you since, for the purpose of your study, such effects are of no practical importance. Alternatively, you may be testing the effect of a new genetic manipulation on increasing the shelf life of apples. In that case, the increased cost of producing the GM apples may mean that to be financially viable, the manipulation must increase shelf life by at least four days, and so you design your study to be able to detect that magnitude of effect. Again, you may well miss smaller effects, but (given your aims) this does not really matter; an increase of only half a day is of no use to you . . . no matter how real that effect is.

In both of these cases, what you have actually done is think more clearly about the actual question you want your study to be able to address. You are not really simply interested in whether the supplement increases dog longevity or GM increases shelf life. You actually want to know whether dog longevity is improved by more than 5 per cent or shelf life by more than four days, and so you can design your study to make sure you have a good chance of answering these questions.

Decisions on effect sizes can be based on many different factors. You might be testing some biological theory that predicts a particular size of effect, and so you need to design your study so that it has a good chance of being able to detect an effect if it is of that magnitude. Another approach that is often used is to look at the sorts of effects that have been seen in similar kinds of studies, and make sure that your study will also be able to detect similar sizes of effect. However, this approach is not without problems. Published effect sizes are likely to be biased upwards (especially if previous studies had low power, as we discussed in Chapter 2); and anyway, who is to say that just because an effect of a certain size has been detected in another study, that makes it biologically relevant for your study? We would advise against this approach, or only using it with caution.

The reality is that there will always be effects that are real, but too small for your study to realistically detect, and so the question for you is how small these need to be before they do not matter for the conclusions you want to draw. We wish there was a simple answer we could give you . . . sadly there is not, you will need to think it through for yourself. But your study will be so much better for having done so!

4.1.2 Inherent variation

We discussed this in Chapter 3, but here is a quick refresher. Inherent variation is a measure of the differences we expect to see between different experimental units in our sample for the variable we are measuring that have nothing to do with any of the experimental manipulations we have carried out. If that is still too opaque, consider the following example. You are measuring the titre of oestrogen in experimental rats. Even within an experimental treatment group, it is unlikely that all rats will have exactly the same circulating hormone level. And even if they did (and we really want to emphasize that for most biological material this will never be the case, even for genetically identical rats), our experimental techniques are unlikely to be able to measure the level perfectly, leading to variation amongst our rats in their measured hormone levels. It is this variation between experimental units that we need to estimate if we want to calculate power. For the purpose of this chapter, we will assume that the variation for our measurement of choice follows a normal distribution, and so can be described by the standard deviation of that distribution. In later chapters, we will introduce you to other statistical distributions that you might sometimes want to use rather than a normal distribution.

OK, so you need to come up with a plausible estimate of how much variation you expect to see in your study. There are many places that this might come from. If your study is part of a wider programme, then it is highly likely that there will be information from previous experiments that you can use to inform your estimate. Even if it is not, you are unlikely to be the first person who has ever worked on this study system (or one very similar), and in that case a quick literature search may well provide you with the estimates that you need. However, we again caution you that studies with low estimates of inherent variation are probably more likely to be published, and thus published values might tend to be underestimates. Further, like anything else, the amount of variation may itself vary from study to study, and you need to keep this in mind when choosing the value you will use. Again, your (or your supervisor's and lab mate's) biological knowledge can help. But what if you find yourself out on the cutting edge, a pioneering researcher working with a brand new study system? In this case, your only solution may be to carry out a small pilot study and estimate the variability directly. However, in this situation, the gods of statistics are on your side . . . it turns out that you only need to measure a relatively small number (seven or eight) subjects to get a pretty good estimate of variation. And this may represent a small additional investment of time, money, and biological material if it means your study has been sensibly designed. If you have read this far, then we know you are someone who will see the value of that investment.

4.1.3 Introducing our example

Let's apply the ideas we have been discussing to a specific example. Imagine you are planning an experiment to determine whether allowing mice more opportunity to exercise by providing a running wheel affects the circulating levels of a stress hormone.

The first step is to decide on your design. The head of the lab suggests that you use 20 female mice, each housed individually in cages, on the basis that this is the number they have always used in the past. You will add a running wheel to half of these cages chosen at random, whilst the remaining cages will be the same in every respect, except that the running wheel will be locked in place so that the mouse cannot use it for exercise. This is what statisticians refer to as a single factor randomized and balanced experimental design. After two weeks in the experimental conditions, you will take a blood sample from each mouse and assay the levels of hormone to give a single measure for each mouse. You will analyse the data using a t-test, which is an appropriate method for this kind of design, and for which the null hypothesis is that providing mice with wheels does not affect their circulating levels of hormone after two weeks. So far so good, you have a clear idea of what you are planning to do.

The next step is to think about the effect size you want to be able to detect (remember, no matter what you do, there will always be an effect size that is too small to be reliably detected by any study). For the purpose of this example, let's assume that the average level of the hormone in mice in typical lab cages is 100 pg/ml of blood. Work by your behavioural colleagues shows that reductions in the stress hormone of less than 10 per cent reflect changes in stress that have negligible impacts on mouse welfare. If your study can only reliably detect changes of 30 pg/ml or above, this means you would be very likely to miss smaller effects that could still have welfare implications. So what about setting a much smaller effect size (say 2 pg/ml)? This would certainly give you a very good chance of detecting effects of biological relevance, but of course would mean you could also detect effects that were too small to be of relevance to your study. At first, the costs of this may not be obvious to you, but think again about what we have said about effect sizes and power. All things being equal, the smaller the effect size you want to detect, the lower the power of a particular planned study. So if you design an experiment to detect a small effect size, it probably means conducting a bigger experiment than would be needed to detect a larger effect. And that means if you design an experiment to be able to detect effects that are smaller than you really care about, your study is bigger than it needs to be. Based on this logic, you decide that you will design a study to be able to detect changes of 10 pg/ml.

The final step is to think about the variation. How much mouse-to-mouse variation in hormone levels do you expect to see? Again, you decide to make use of previous studies performed in your lab that show that the typical mouse-to-mouse variation you see for the measurement you are planning to take follows a normal distribution, and that a reasonable estimate of the standard deviation of this distribution is 10 pg/ml.

 Key point

In order to calculate the power of a potential experiment, you need to fully describe the experiment and estimate the effect size of interest and the inherent variation that you will likely have to cope with.

4.2   The concept of estimating power by repeated evaluation of synthetic data

OK, so you have done the hard thinking about the things that you need in order to estimate power, but how do you actually use this information? The traditional answer is that you plug it into an online power calculator or statistics package, and your power value pops out. This is all well and good if you are using a simple design for which a power calculator is available. However, if you plan to stick in research for any length of time, then before long you will probably find yourself planning a study for which this is not the case. For example, it would be easy to find an online calculator for your mouse wheel study if you plan to have equal numbers of subjects in your two groups. However, what if you are considering a design with unequal sample sizes for your treatment groups (later on we will suggest some reasons why this might be worth considering)? With unequal sample sizes, you might find that off-the-shelf power calculators soon start to struggle. Our aim here is to teach you a different, more general approach to estimating power. Our approach will work for simple designs, but has the great advantage that it can be extended to any kind of study that you might be able imagine. (We think that there are added advantages too: your understanding of the experiment should deepen from calculating power our way, you don't have to learn how to use a new package if you are already doing statistics in R, and you don't have to pay for a new package.)

The approach is conceptually very simple: we generate an imaginary data set of the sort that we plan to collect in our actual study, and we then analyse this data to see whether we can detect the effect we are interested in. The trick is that we use the assumptions we have made about variation in our biological system and the effect size that we are interested in when we generate our data. You can think of this as carrying out an experiment in an imaginary world where the variation mimics your understanding of variability in your real biological system, the null hypothesis is definitely false, and the size of the effect has been set to a value you want to be able to detect. If, based on the results of our statistical test, we fail to reject the null hypothesis (even though we know it is false), we have made a Type II error. We then repeat this procedure many times, each time generating a new imaginary data set, and each time seeing whether we successfully reject the null hypothesis or make a Type II error. That is, we conduct the same imaginary experiment but drawing a different sample randomly from the underlying distribution that we assume for the population. Suppose we do this 100 times, and make Type II errors in 15 of our imaginary experiments, then our estimate of the Type II error rate is 15 per cent, and so our estimated power is 85 per cent. In reality, we would carry out way more than 100 imaginary experiments to give us a more precise estimate, but hopefully you get the idea.

Now the description above might seem a bit too abstract to you, so let's now actually go through an example of the process.

4.2.1 A concrete example

In Chapter 1 we introduced the idea that an experiment can be thought of as drawing samples from populations. So our first step is to define the populations in our study. The hypothetical populations we will use in our mouse study are shown in Fig. 4.1. Both populations follow a normal distribution with the same standard deviation of 10.

Fig. 4.1 The normal distributions from which we can draw our simulated data sets.

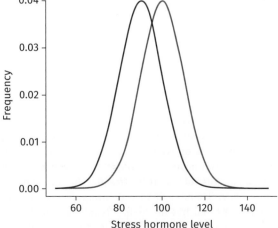

However, because we are interested in a world where the null hypothesis is false, and wheels do affect the level of stress hormone, the two populations have different means. The actual standard deviations and means are based on the assumptions we made above.

To carry out our imaginary experiment, we need to generate individual hormone measures for our 10 control mice and 10 experimental mice. This is easy to do with a random number generator. First we draw 10 random numbers from a normal distribution with mean 100 and standard deviation 10. These are the measures from our 10 control mice. We then repeat the process drawing 10 numbers from a normal distribution with mean 90 and standard deviation 10 to give us the measures for our mice that can exercise on their wheels. This procedure generates exactly the kind of data we expect to collect in our planned study with real mice, and one example of such a synthetic data set is shown in Fig. 4.2.

However, with our imaginary data we have one big advantage over the real world: we know what the correct answer is. The null hypothesis (that the mean stress hormone levels in the two groups are the same) is false, because we have set things up that way. For a minute though, let's forget we know that, and continue in the same way as we would do with our real study. The next step with our real data would be to carry out a statistical analysis to see whether any differences we see between our samples provide evidence to suggest that the null hypothesis is false.

Having planned our study meticulously, we already know that the way we will analyse our real data is to use a t-test to compare the means of our two samples. So let's do the same with our simulated data. A few key-strokes in an appropriate statistical package gives us our result: the p-value associated with our statistical test is 0.017. Since this p-value is below the threshold we have decided on for statistical significance (0.05), our conclusion is that our data provides enough evidence for us to reject the null hypothesis. Put another way, our imaginary experiment has given us the correct answer (or using the statistical speak we introduced in Chapter 1, we have not made a Type II error). At this point, we might be tempted to let out a small cheer. However, before you celebrate too much, we are not finished yet. Remember, our result depends on the particular set of random numbers that were generated in our imaginary experiment (i.e. the particular sample that we drew from the underlying population). If we repeated the experiment, we would expect to get different numbers (i.e. a different sample), which might cause us to draw different conclusions. To estimate how often we would expect to get the right answer, we need to run a number of simulated experiments and then work out the proportion of simulated experiments that would correctly reject the null hypothesis. Fortunately, computers can do these sorts of things very quickly, and with some simple programming we can get the computer to repeat the process for us, and provide

Fig. 4.2 An example of a simulated data set for our mouse experiment with ten mice in each group.

```
> control
 [1]  120.96529     91.80591    117.37809    109.46367    97.75557    114.60053    97.94677
 [8]   90.67640     96.18378    103.58138
> wheel
 [1]   77.47319     88.29145    105.17283     80.38739    81.79582     83.68925    96.84645
 [8]   77.97649    109.89127    102.71595
```

Fig. 4.3 The distribution of p-values obtained from 100,000 runs of our imaginary experiment.

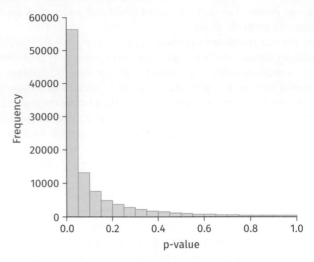

us with all of the p-values from all of the simulated studies. So let's repeat our study 100,000 times and have a look at the results. The distribution of the p-values from these imaginary experiments is shown in Fig. 4.3.

The first thing to notice is that we don't always get the right answer. Even though we know the null hypothesis is false (that is, our experimental and control samples are being drawn from normal distributions with different means), sometimes the difference between our samples is not great enough to generate a significant p-value with our statistical test. In all of the imaginary experiments where the p-value from the t-test is greater than 0.05, we are making a Type II error. In all of the other experiments, we get the right answer, and correctly reject the null hypothesis. However, we are scientists and just contemplating the picture is not enough. We need to make things more quantitative. To do so, all we need to do is count up the number of experiments where we got the correct answer (of course we actually get a computer to count for us and in this case that comes to 56,275). So out of 100,000 imaginary experiments in which the null hypothesis was false, and the ability to exercise in a wheel reduced the level of stress hormone levels by 10 per cent, our planned experiment was able to correctly reject the null hypothesis in 56,275 of them. Or put another way, the power of our planned study to detect an effect of 10 pg/ml is 0.56, which means we would only get the right answer just over half of the time. Oh dear, that's not as high as we might have hoped!

 Key point

We can calculate the power of our potential experiment by drawing repeated samples from a population that we have set up with our assumptions about inherent variability and effect size. The estimated power in this scenario is simply the fraction of the repeats that yield a sample that cause us to correctly reject the null hypothesis. If we have lots of repeats, our estimated power will be close to the actual power.

4.3 The nuts and bolts of generating synthetic data and estimating power in a single one-factor design

OK, now we have been through the process and seen how it works. But the point of this book is to teach you how to do it yourself, so now it is your turn to write the simulations. We assume that you have already installed RStudio on your computer. We also assume that you know how to open a script window, edit, run and save scripts in RStudio. If not, take a break from the book and go and work through our very very basic guide to RStudio, which can be found with the online material associated with this book and we will see you back here in a few hours.

Start RStudio and open a new Script window. You should see a new window in the top left quarter of your screen, with something like 'Untitled1' in the tab at the top. A flashing cursor awaits your commands (if the cursor is not flashing, click in the script window and it should start). The first thing you should do is add a comment at the top of the script to remind you what the script is for. Type the following at the cursor and then press return:

```
# Script to generate data for a one-factor design
```

Remember, '#' at the start of a line tells R to ignore everything else that follows on the line, so this will just be a comment to remind you of what follows. We cannot emphasize enough the importance of including lots of comments in your scripts. What may seem obvious and unforgettable to you now will often quickly be forgotten, and the time spent deciphering poorly annotated scripts can quickly move from minutes to hours to days!

Let's start by generating our ten control mice using a random number generator. This is easy in R, as it has a built-in function to generate random numbers from a normal distribution (and lots of other distributions . . . more on that later). All you need to do is tell R how many random numbers you want, and what the mean and standard deviation (sd) of your normal distribution are. So to generate our 10 control mice, all we need to do is add the following line to our script:

```
control <- rnorm(10, mean = 100, sd = 10)
```

What does this mean? rnorm is the name of the R function that generates random numbers drawn from a normal distribution. The information in the brackets tells R how many random numbers we want it to generate (in our case ten, but we could change this to any number we want) as well as the mean and standard deviation we want our normal distribution to have. Computer programmers call these the arguments of the function. The first part of the line tells R to take the vector of ten random numbers that it has generated and give the vector the name 'control'. Obviously, you could use any name you like.

However, none of that will happen until you run the line, so let's do that now. Click your cursor somewhere on the line in the script window, and then click on the run button at the top of the window. All being well, R has run your line of code, and your control mouse values have been generated. However, you are probably feeling justifiably underwhelmed at this point as at first sight it is not obvious that anything much has happened. If you look at the Console window, you should find your line of code has magically appeared there as well (and hopefully there are no error messages. If there are, check your typing carefully and try again). If you are particularly observant, you may also notice that the Environment window in the top right quarter of our screen contains some new text.

```
control          num [1 : 10]  106.1 88.4 86.3 97.7 93 . . .
```

The Environment window is where RStudio shows us the things we have asked R to remember, and so the new text tells us R now knows about a vector called control, which contains ten numbers, and it also gives us the first five numbers in the vector (these will differ from the numbers you get, remember it is random). So this is looking hopeful. However, to actually see our 'mice' in all of their glory, we need to add an additional line of code. Go back to the script window and type the following on the next line:

```
control
```

and run it. Remember, we asked R to store the vector of mice we generated and give it a name, 'control'. Typing 'control' tells R to display that vector (i.e. to show us 'control'). If you now look at the Console, you should see that this new command has been pasted into the Console, and below it should be a list which, to the uninitiated, just looks like ten numbers. To us, however, these are the ten measurements from our control mice. In general, as you build up your own scripts, it is good practice to check it is doing what you expect by outputting everything (even if you go back and remove these outputs from the final version).

We can now do exactly the same for our wheel treatment mice, remembering that we need to change the mean to reflect the fact that the wheel treatment has an effect on hormone levels. You can probably make a good guess at what that code might look like, but here it is:

```
wheel <- rnorm(10, mean = 90, sd = 10)
wheel
```

As before, type this in and run it to check that it does what you think. You can either run the lines one at a time as above, or if you want to be really flash you can use your mouse to select both lines at once.

If all has gone to plan, you have now carried out your first imaginary experiment, and have hormone measurements from ten mice from the control population and ten mice from the population of mice with wheels. Even if you have never programmed a computer in your life, you have to admit, that was considerably easier than carrying out the two-week experiment with 20 real mice! The final stage is to analyse these data in the same way as you would with your real data, and see whether you can successfully reject the null hypothesis (remember, we know it is not true. The question is: can we detect that with the imaginary experiment we have carried out?). We had planned to analyse our data using a Student's t-test, and the code to carry out such a test on our imaginary data is as follows:

```
t.test(control, wheel, var.equal = TRUE)
```

Unsurprisingly, the t.test function is R's go-to function for carrying out a t-test. If you want to find out more about the t.test function, or any other function you know the name of, you can make use of R's extensive help documentation. To do this, go to the Console window and type ?t.test then press return. All you need to know here is that the bits in brackets tell R the names of the two vectors containing the data to compare, and whether we want our t-test to assume equal variances or not. The traditional Student's t-test assumes that both groups being compared have equal variances. Add this line to the bottom of your script and run it; you should see the output of the t-test has appeared in your Console window. Have a look at your results. Assuming a typical significance threshold of p = 0.05, would you reject or fail to reject the null hypothesis on the basis of these data? Remember, we know the null hypothesis is false, so this amounts to asking whether you have got the right answer or not. So were you right or wrong?

Congratulations, you have now simulated your first imaginary experiment and analysed the results. This would be a good time to go back to your script and add lots of comments (remember #) to remind 'future you' of what you have done. It would also be a good time to save your script.

Try running the whole script again. You will see that the results have changed. The t-value and p-value will be different. This is not surprising, even though our populations remain the same (and the null hypothesis remains false); the particular random samples we draw will differ each time. Sometimes we will reject the null hypothesis, whilst other times we will not. To say anything useful about the power of our planned study, we need a good estimate of how often we get the right answer, and that means running lots of imaginary experiments. Fortunately, now that we have a script to run one experiment, it is reasonably straightforward to expand it to automatically run the experiment as many times as we like. To do so, we need some new programming tricks.

The first is something called a loop. A loop is just a way of making a computer program repeat the same operation over and over again. To avoid messing up our current script, open a new script window and enter the following code:

```
for(i in 1:10) {
print("Hello")
}
```

A basic loop has three parts. The first line opens the loop with a curly bracket, and tells R how many times we will go around. In this case, our loop will count up from one to ten, giving us ten iterations. The final line closes the loop with a curly bracket. Finally, the middle part contains the lines of script that will be run each time we go round the loop. In this case, our script is simply a single line of code that uses the print command, which is an R commend that prints whatever is inside the brackets to the Console. This script is only a single command, but script inside loops can be as long and as complicated as we like, and we will make use of this fact later on. Run the script and have a look at the Console. You will see 'Hello' printed ten times. Success, we have automated our script to run multiple times. Changing the number of times we run the program is as simple as changing the numbers in the first line of the loop. If we change ten to 100, then our program will run 100 times (don't take our word for it . . . give it a try!). What this means is that, in principle, we could replace our simple salutation script with the imaginary experiment script we wrote earlier and automatically generate thousands of p-values. However, there are a couple of other issues we need to deal with before that. The most obvious is that even if we automate the program to generate and analyse the experiments, we would still be left with thousands of t-test results to work through. This idea should fill you with horror. Fortunately, we can get R to analyse the outcomes of all the t-tests for us.

Go back to your imaginary experiment script. The first thing we are going to do is edit the line that carries out the t-test to read as follows.

```
my.result <- t.test(control, wheel, var.equal = TRUE)
```

What this does is run the test as before, but instead of sending the results to the Console, it saves the results of the t-test as what R calls an object and gives the object the name 'my.result'. If you don't believe us, add the following additional (and very simple!) line of code:

```
my.result
```

If you run this line, R will send the object you have called 'my.result' to the Console, and you should see the result of your t-test.

Why do we need to save our result as an object? The t-test results contain lots of information, but for our current purpose, all we actually care about is the p-value. By saving the output as an object, we can then use other R functions to pull out the bit we need. So if we now enter and run the following code:

```
my.result[3]
```

R sends the third component of the object called result (remember result is the entire output from the t-test) to the Console, and this third component just happens to be the p-value for our test. How did we know that the p-value was the third component, not the second or fourth? This is not the place to go into that, but we will return to it in Chapter 5. However, for the time being, just take our word for it.

Rather than sending the p-value to the Console, let's save the value and give it the label 'pvalue', just like we did above for our mouse data. The code is:

```
pvalue <- my.result[3]
```

We are almost there. All we need now is a way for R to decide for each test whether the p-value is significant or not, and keep track of the number that are. And to do that, we will make use of two final coding tricks: counters and conditional operators!

The idea of a counter is very simple: we define a label in R and give it an initial value of 0. Then, every time something we want to count happens in our script, we add 1 to the current value of the label. To see this in action, go back to the script window where we were playing with loops earlier. We are going to add a counter that counts how many times we go around the loop. Edit the script until it reads as follows (new lines are in bold):

```
count <- 0
for(i in 1:10) {
print("Hello")
count = count + 1
print(count)
}
```

The first line gives our counter a label (imaginatively we have gone for 'count'. . . You could choose 'Nick' or 'Graeme' or anything else; it doesn't matter, it is just a label), and sets it to its initial value of 0. The next line is inside the loop, and so forms part of the script that will run each time we go round the loop. It tells R to give count a new value, which is equal to the current value plus 1. To show that it is working, we have added another line to our script inside the loop that again uses the print command. Now the observant amongst you will notice a subtle, but important difference between the two print commands. In the first print command, the word 'Hello' in the brackets is contained within speech marks. In that case, R prints the actual word (and its speech marks). The second print statement does not use speech marks, and so rather than printing the word 'count', R looks for something with the label 'count', and prints that instead. In your script, count is the variable that stores the current value of our counter, so R prints that value to the Console. So now, our program doesn't just greet you each time it goes around the loop, it adds 1 to its counter, and prints out the current value of the counter as well. Now counting greetings is all very well, but what we want to be able to do is use a counter to count how many of our imaginary experiments generate significant p-values, and to do this, we need to combine our counter with a conditional operator.

Conditional operators are a way of telling a computer to do different things under different circumstances. In our case, what we want to do is get R to see whether our p-value is significant or not. If it is significant, then we add 1 to our counter, but if it is not significant, then we leave the counter at its current value. To do this, we will make use of something called an `if` statement.

Go back to your imaginary experiment script and add the lines in bold:

```
count <- 0
control <- rnorm(10, mean = 100, sd = 10)
control
wheel <- rnorm(10, mean = 90, sd = 10)
wheel
my.result <- t.test(control, wheel, var.equal = TRUE)
pvalue <- result[[3]]
print(paste("pvalue = ", pvalue))
if(pvalue<0.05){count = count+1}
print(paste("count = ",count))
```

The first new line sets up our counter as before. The conditional statement can be found at the bottom starting with the word `if`. R tests whether the condition in brackets is true, in this case, if the p-value from our test is less than the threshold of 0.05. If the condition is met, then R runs the code in the curly brackets. In this case, it is a single line which adds one to the counter, but as with the loop, we could have a much more complex, multiline code in there. If the condition is not met, then R just moves on. We have also added two more print commands to our script that send the p-value from the test and the current value of the counter to the Console. We have added these to allow you to check that the program is doing what you think, and you will probably remove them again later once we are happy everything is working. Again, this print function is slightly different from the previous ones, using another function, paste, to let us print both the words in the speech marks and the value together. We are not going to go into the details of how it does that (this really isn't a programming manual, and remember, you can find out more about specific R functions by either using the R help system as described above or by consulting the internet), but we have included it so that you can use it in other scripts. Now run the script. Was the p-value significant? If so, the counter increased from 0 to 1. If not, the counter stayed at 0. We can now repeat this process by re-running the script. If you do, make sure to not rerun the first line, or you will reset your counter to zero again. Run the script ten or 20 times, and check the counter is working. If it is, you have now written a script that not only generates an imaginary experiment and analyses it, but automatically checks whether the test is significant, and keeps a count of how often that happens. Well done! This would be another good point to make sure you have saved your work and added some (lots of!) clear annotation.

Now we can get back to automating the whole process by putting our script inside a loop. The main thing to be careful of here is to think about where the loop should start and finish. Remember, everything inside the loop's curly brackets will be run each time the loop runs. This means the loop should start after we have set our counter to 0 . . . otherwise the counter will be reset every time we go around the loop which would clearly not be helpful! Similarly, unless we want the program to give us the value of the counter after each imaginary experiment (which would slow the computer down terribly), we probably want the final line of the script to be outside the loop. See if you can add the required lines to put your code into a loop. If you are struggling, look at Script 4.1 below. Or you can download an electronic version of the script from the ESM (Electronic Supplementary Material) associated with this chapter which you can load straight into RStudio.

```
#Script 4.1
#Simulates a single factor study with two levels
#that will be analysed with a t-test.
#This line clears out R (see intro ESM for more details).
rm(list=ls())
#Sets up a counter that we will use to count significant
#results and sets it to zero.
count <- 0
#The start of the loop. The code within the loop is
#repeated 100000 times, generating a data set and P value
#for each iteration. Thus, our estimated power will be
#based on 100000 imaginary experiments.
for(i in 1:100000) {
#Generate the random values for our control mice.
control <- rnorm(10, mean = 100, sd = 10)
#Generate the random values for our treatment mice.
wheel <- rnorm(10, mean = 90, sd = 10)
#Carry out a t-test on our imaginary data, and store the
#output in an object called my.result.
my.result <- t.test(control, wheel, var.equal = TRUE)
#Extract the P value from my.result, and store it in a
#variable called pvalue.
pvalue <- my.result[3]
#Check to see whether our P value is significant.
#If it is, it adds one to the value of the counter.
if(pvalue<0.05){count = count+1}
#The end of our loop.
}
#Once the specified number of imaginary experiments
#have run, send the value of the count variable
#to the Console to give us our answer
print(paste("count =",count))
```

4.3.1 Final checks and tweaks

Estimating power will require you to run thousands of imaginary experiments, and a loop lets you do that easily. However, before rushing into that, we suggest some final checks to make sure everything is working as expected over a small number of runs, so set your loop to run the script ten times, and then run it. We have left the line of code that outputs the p-value for each experiment to the Console. Have a look at the list of p-values and count how many are significant. Does this correspond to the count you get at the end? If so, all is probably well (although it is worth doing this a few times just to be sure). Once you are happy, it is worth reviewing your script. There may be some lines that you included as part of the script development but are no longer necessary. For example, the line to print out the individual p-values was very helpful in making sure the program was working properly, but is not really necessary now that we are happy things are working. If we leave it in, then R will output all of potentially thousands of p-values to the Console, which is unnecessary and actually slows the script down quite substantially. We would recommend you remove it, either by simply deleting it, or by typing # at the beginning of the line. This turns it from a command that R will execute into a comment that R will ignore, and can be a very useful trick if you think you might need to put it back in again if you want to develop the script further.

You have finally made your assumptions, written and tested your script, and you are ready to estimate the power of your planned study. Go to your script and edit your loop so that it will run 100,000 imaginary experiments, which should be enough to give you a reasonably precise estimate of your power. Select the entire script, and run it. The first thing you will notice is that, unless your computer is considerably faster than ours, nothing much happens for a while. The script has been copied to the Console, but you don't have a final count. This is because, even for a computer, 100,000 imaginary experiments take a while to do. However, eventually the computer finishes its work and tells you how many of your 100,000 imaginary experiments gave a significant result. In our simulation we got 56,330, and hopefully you will have something similar. Now remember, the power is defined as the probability of correctly rejecting the null hypothesis given that it is false. The number of imaginary experiments where we correctly rejected the false null hypothesis was 56,330, so our estimated power is 56,330/100,000, or 0.5633.

When you run the script, you will not get exactly the same value as us, because your computer's random number generator will select different samples. However, providing we run enough replicates (10,000 or more as a rule of thumb), the estimates we get out for our estimated power should be pretty repeatable. If you want to read a bit more on this, then we have some more material in the ESM for this chapter. The more replicates you run, the longer the computer will take. Throughout this book, we have written R code in a way that we hope is as easy to follow as possible. But there are some hints and tips for ways you can write code a little differently in order to speed it up, and we have some material for that in the ESM for this chapter too.

 Key point

Turning your conceptual understanding of calculating power by simulation into actual power values is simple in R.

4.4   Using simulation to compare alternative ways of doing the same experiment

Now that you have your basic simulation up and running, it is reasonably straightforward to modify it to consider how changes to the planned design affect its power. The most obvious aspect you might be interested in is how the number of experimental subjects you use affects the power. We discussed in the last chapter why there are a suite of other factors you should probably think about first, but let's run with sample size for now. We also learned in the last chapter that increasing sample size should increase power; but our simulation allows us to explore this relationship. That is, we get to quantify how much a specific increase in sample size would buy us in terms of increased power. This can be extremely useful, for example, if you find your planned study is not powerful enough and want to know how much bigger it would need to be to give you the desired power. Alternatively, if you find your planned study already has very high power, you might decide to see whether you could use fewer individuals without reducing the power too much.

Let's start by doubling our sample size to 20 mice in each group. To do this, simply get the program to draw samples of 20 instead of ten. Remember the line in the script that generated your control sample:

```
control <- rnorm(10, mean = 100, sd = 10)
```

To increase our sample to 20, simply edit this line replacing the ten with 20. We can then do the same for the equivalent line for the treated mice. Having made these changes, we can re-run our simulation and estimate power for our new design. We can repeat this for any sample size we are interested in. As you become a more confident programmer, you will find it is also possible to automate this procedure, effectively using loops to repeat the entire process for different sample sizes (you can find an example of how to do this in the ESM associated with this chapter). However, this will get us where we want to go, if a little slowly, so let's not try to run before we can walk. Try running your simulation for 5, 10, 15, 20, 25 and 30 mice per treatment and noting down the power in each case. Perhaps you want to plot the data? We have, and the plot is in Fig. 4.4.

Fig. 4.4 Estimated power in our mouse experiment as a function of the number of mice we might decide to have in each of the two groups.

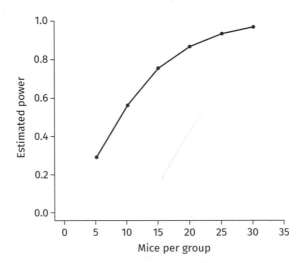

Your planned design has equal numbers of mice in each treatment. There are often good reasons to keep your design balanced like this, but under some circumstances there might be arguments to consider a design with differing numbers of individuals in each group. For example, if the treatment is unpleasant for experimental subjects, whilst the control is relatively benign, we might want to know what would be the effect on power of having a design in which we had fewer individuals in our treatment group than in our control group. We can easily answer this question with our simulation, again by adjusting the number of samples drawn for our treated and control groups. Try using the simulation to compare a balanced design with 20 individuals in each group, to a design with 25 control individuals and 15 treated. What effect does this have on power?

You can also explore how changing other aspects of the study might affect power. In your mouse wheel example, you planned to measure the circulating levels of a particular stress hormone. However, there might be other measures you could take. Suppose you are also considering an alternative stress measure that you might use instead. You know that this measure shows less variation amongst individuals, which should enhance your power. However, your behavioural colleagues suggest that mice are more sensitive to changes in this second measure. As a result, you would want to be able to detect smaller effects with this second measure, which would reduce the power of your study. Intuition is a poor guide to which of these effects would be most important, but we can easily change the assumptions we made about the populations to fit this new measure. For example, say the new measure has an average in controls of 100 pg/ml with an sd of 5, and we want to be able to measure an effect of the treatment of reducing this to 94 pg/ml. Try running this design and see what effect it has on power.

So far, we have focused on a simple one-factor design, as this was appropriate to address our particular research question (and made for a simple first case to learn the basics). However, for other more complex questions, there may be several different designs that, in principle, could be used. The great strength of the simulation approach you are developing is that it can be applied in turn to each of the different designs, allowing you to make sensible decisions about the best way to answer your research question. In the next few chapters, we will start to look at some of the techniques you can use to simulate more complex studies.

 Key point

Once you have written code to estimate power for one possible version of your experiment, it is very easy to modify it to compare different versions, to help you pick the one you feel delivers high power most efficiently.

4.5   Selecting between different alternative experiments to decide which experiment you are actually going to do, and reporting its power

OK, you have written a script that allows you to simulate any one-factor study you can imagine and you have used it to explore how changing aspects of the design affects its power. And you have learned a number of basic scripting tricks along the way, which will allow you to simulate more complex designs in the future. However, this is not the end of the story; we still have one final thing to do, which is to interpret this information and decide which study to actually carry out. So, let's finish this chapter by thinking about how you might do that.

Let's start by thinking about the figure we showed earlier relating the power to the sample size. How do you use this to decide which sample size to choose? When people first start to think about power analysis, a common mistake they make is to assume that you always want the most powerful study. However, if things were as simple as that, we wouldn't really need power analysis. . . The biggest experiment is always the most powerful, so we would just do the biggest that we possibly can. So what is wrong with this argument? Well the first thing to remember is that experiments are not free. Every time we add additional replicates to our study, this comes with additional costs in terms of money, additional time and if our work involves animals, ethical costs. If our study already has high power, then these additional costs will give us a relatively small increase in power. So in our figure, going from 20 mice per group to 30 mice per group represents a reasonably large increase in the number of animals . . . but our power increases by less than 0.1. It is hard to imagine that the increased power would justify the costs. However, at the other end of the scale, going from 5 mice to 15 increases power by 0.47. So the same number of extra mice leads to a substantial increase in power, and could be more easily justified. Based on this logic, there should be some optimum level of power, above which our experiment is too big, and below which it is too small. The figure that is often adopted as a threshold is a power of 0.8 or 80 per cent, and adopting this in your study would satisfy most people. Much below this, you should question whether your study is worth doing in the first place, and much above you should question whether you could save resources with a smaller study. In our mouse wheel experiment, it looks like a study of about 17 mice per group would be close to the optimum.

Our decisions may also need to go beyond the simple consideration of total sample size as well. We discussed earlier how we might use the simulations to examine the effects on power of using an unbalanced design, in order to reduce the number of individuals that experience an unpleasant treatment. Now suppose that we find that two designs have a power of 0.8. The first uses a balanced design with 15 mice in each group, the second is unbalanced with ten mice in the treatment group and 30 in the control. So the second design uses more mice in total, but fewer in the unpleasant treatment. Which of these you decide to use will depend on a number of biological factors that are specific to your system. However, the explicit consideration of power tells you which possible designs you should be deciding between. Try running these values (and any others that you fancy), and see what you conclude. We consider some of the other decisions you might make based on your simulations in section 4.6.

Hopefully you are now starting to think that this simulation approach is not so difficult, and even imagining ways that you could extend the approach to other kinds of designs (go ahead and play with the script, change things and see what happens—you won't break anything, and playing is the best way to learn!). In the next chapter, you will build on what you have learned, and think about some more complex designs. You will also learn some additional scripting tricks that will make your scripts quicker and easier to modify.

Case study 4.6

Presenting and interpreting your power analysis

In the main text, we have focused heavily on the effects of sample size on power. This is for the simple reason that considerations of sample size were, traditionally, the most common use of power analysis. Sample size is also often one aspect of the study that the researcher can easily alter. In this case study, we will think in more detail about some of the other decisions that power analysis can help us to make. To do so, we are going to imagine that we are writing a grant to seek funding for the mouse wheel study. As part of our grant, we will want to show readers that we have properly considered how different aspects of our study affect its power when deciding on the final design. In preparation, we have run our simulation for a variety of different sample sizes, with different assumptions about the inherent variation and the effect of the treatment. The challenge here is to present a lot of information in a compact but still easily intelligible way. On one hand, we think some readers of your grant proposal will be interested in the details of your power analyses. On the other hand, these analyses are unlikely to

be the most important aspect of your document. The most important aspects of the document to the overwhelming majority of readers will relate to your research question, not to the methodology of your work (like your power analysis). Thus, painful as it is for us to admit, the discussion of your power analysis will probably take up only a small proportion of documents that you will produce for others to read. This is not an argument for not presenting your power analysis, because some readers will not be willing to simply take on trust that you have done a thorough job. It is, however, an argument for trying to present it in as compact a fashion as possible: we think that, ordinarily, what you are aiming for is a single (often multi-panelled) figure, with a very full caption. Let us illustrate this by example: you might end up describing your power analysis in something like the form shown in Fig. 4.5. Such a figure allows you to justify aspects of the experimental design, whilst showing the robustness of the chosen design to any departure from your best guess for effect size and inherent variation.

Fig. 4.5 The estimated power of 100,000 simulated mouse wheel experiments with sample sizes between five and 20 mice per group to detect a significant difference in stress hormone levels. In Panel (a) we explore the effect of varying our estimate of mouse-to-mouse variation on our power to detect an effect size of 10. Based on our best estimate of an sd of 10, our power analysis suggests using at least 17 mice per treatment to achieve our desired power of 0.8. With a higher sd of 11, the power at this level drops to 0.730 and so our design is reasonably robust to imprecision in our estimate of the true sd. Panel (b) explores the effects of changes to the effect size on the power of our study (assuming an sd of 10). Power is strongly affected by effect size, and we do not feel we could justify the sample sizes required to obtain good power to detect effect sizes below 10.

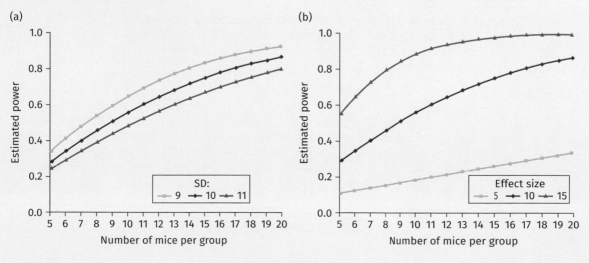

(a)

(b)

 Summary points

- In order to calculate the power of a potential experiment, you need to fully describe the experiment and estimate the effect size of interest and the inherent variation that you will probably have to cope with.
- We can calculate power for our potential experiment by drawing repeated samples from a population that we have set up with our assumptions about inherent variability and effect size. The estimated power in this scenario is simply the fraction of the repeats that yield a sample that cause us to correctly reject the null hypothesis. If we have lots of repeats, our estimated power will be close to the actual power.
- Turning your conceptual understanding of calculating power by simulation into actual power values is simple in R.
- Once you have written code to estimate power for one possible version of your experiment, it is very easy to modify it to compare different versions, to help you pick the one you feel delivers high power most efficiently.

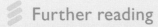 Further reading

Some books we like on writing your own programs in R:

Grolemund, G., 2014. *Hands-On Programming with R: Write Your Own Functions and Simulations*. O'Reilly Media, Inc.

Templ, M., 2016. *Simulation for Data Science with R*. Packt Publishing Ltd.

But there are lots of such books; you should browse a few to find one that chimes with you.

Discussion questions

4.1 Are there ever any circumstances where any non-zero effect you detect (no matter how small) would lead to very different biological conclusions from detecting no effect?

4.2 Discuss the decision to only use female mice in our exercise wheel and stress hormone experiment.

4.3 We argued that you might opt for an unbalanced design if one of your treatments was considerably more stressful to animal subjects than the other—can you think of other circumstances where you might adopt such a design?

5 SIMPLE FACTORIAL DESIGNS

Learning objectives

By the end of the chapter, you should understand:

- How you would organize the data from your experiment in a spreadsheet and use that as a guide to producing exactly analogous simulated data.

- That if you can produce simulated data organized in the same way as your actual experimental data, then conceptually (and practically) calculating power is just as easy for more complicated designs as it was for the simple designs of the previous chapter.

- That once you have the code to calculate power for one design of your experiment, it often requires very little extra effort to modify it to allow you to compare alternative designs.

- How we can usefully use power analyses to explore the effects of adverse events leading to unplanned loss of some of our data.

Chapter 4 covered the general principles of using imaginary experiments as a way to estimate the power of your planned study. Over the next few chapters, we will extend these ideas to studies with more complex designs. By the end of these chapters, we hope that you will see how, with a little thought and only basic programming skills, you will be able to simulate any experiment that you might be thinking of carrying out. Here we begin to complicate things very slightly from the type of experiments described in Chapter 4, by allowing you to have more than one factor in your design.

5.1   Introducing our focal example

In this chapter, we will focus on the following hypothetical study. We are interested in whether resistance to herbicides comes at a cost to plants that would be paid if they grew in environments without herbicides. To explore this question, we will make use of two genotypes of *Chlamydomonas reinhardtii*, a single-celled algae that is a convenient lab system to explore such questions. One of these genotypes, let's call it *Resistant*, has been genetically modified so that it can grow in the presence of glyphosate (the herbicide better known as Roundup). The second genotype is the wild-type. It is killed by a sufficiently high dose of glyphosate, so we call this genotype *Sensitive*. To address our research question, we are going to measure the growth rate of the two genotypes in the absence of glyphosate. So far, this study has just one factor (genotype, with two levels). So what's new here?

The first complexity involves the set-up of our laboratory. Growth rates will be measured in test-tubes of liquid growth medium, and these tubes will be kept in a controlled environment growth chamber where fluctuations in temperature and lighting can be minimized. The chamber that we have available contains three shelves, each of which has space for 30 test-tubes. This means that the maximum size of our growth rate assay is 90 tubes spread across the three shelves. Now, conditions inside our growth chamber are controlled, but they are not perfectly uniform. We know from previous studies that there is a slight temperature gradient such that algae grow slightly faster on the top shelf, and slower on the bottom shelf, than algae in the middle shelf. So the first question we will focus on is: how should we allocate the replicate tubes of our two genotypes to the three shelves? One option would be to use a completely randomized design. Assuming that we are planning to have equal numbers of the two genotypes (and for the reasons we saw in the previous chapter this is usually a good idea), this would simply involve randomly allocating 30 tubes to each shelf, ignoring the genotype. An alternative we might consider is a randomized block design, where we randomly allocate the 45 tubes of each genotype separately such that we always get 15 of each on each of the three shelves. Which of these will be best? To answer this we can simulate these two designs and compare their power directly.

The second complexity is that *C. reinhardtii* cultures are prone to contamination by yeasts and bacteria that float around the lab. This means we expect some of our test-tubes to become contaminated and so provide no useful data for our study. So our second question is: how will this loss of replicates due to contamination impact our power? Again, we can add this random loss to our simulations and explore its effects quantitatively.

 Key point

Additional complications in your experiment can arise for practical reasons rather than solely from the questions you want to explore.

5.2   Think about your data frame as a way to envisage your experiment

A good first step in planning our simulation is to think about the kind of data frame that we need to produce for our planned analysis. In the previous chapter we could ignore this, because to do a t-test in R you don't need a data frame, it can be done directly on the two lists of measurements. However, for most other types of analysis this is not possible and so we need to give a little thought to how the data that we collect will be organized so that R can analyse it for us. For many types of analysis, including the one we will use for this study, the same general type of data frame is used. This is great news for us, as it means that principles we learn here can be applied to a wide variety of studies.

Now we obviously can't do an analysis without data, so our data frame will need columns for each of the different responses in our study. For the herbicide study, we have only one response variable, the growth rate measure from each test-tube, and this column will ultimately contain the growth rate data that we observe. For more complex studies we might measure several traits on each experimental subject, and in that case we would need a separate column for each such variable in our data frame. The other thing our data frame will need are columns that tell R the values of any predictors associated with each measure of the response variable. Here we need one column that contains a code telling R whether a particular growth measure is from a test-tube of the sensitive or resistant genotype, and a second column that contains a code telling R which shelf the test-tube was on. Of course, more complex studies might contain additional factors, and in that case our data frame would need a column for each of those as well. If you are still struggling with getting your head around this, we have put the data frame we are after in Fig. 5.1 (to save space, we have only shown the top few lines of the data frame).

Really, the key to seeing how to generate simulated data is to imagine how the data for your experiment might look on a spreadsheet. You would then likely be inputting your data from the spreadsheet into a data frame in R. Your simulated data is going to have to be of the same form as your real data: that is, in an exactly analogous data frame of the same size and form.

 Key point

Imagine how the data from your experiment will look in a spreadsheet; your simulated data will have to be in exactly the same form.

Fig. 5.1 The first ten lines of the data frame we will produce for our experiment. Each row corresponds to a separate test-tube in our experiment. The first column contains a unique code for each row that R will use to identify the test-tube that the row corresponds to. In this case, the first ten entries are all for the *Sensitive* strain. The second column contains a code corresponding to the shelf the tube is growing on. The final column contains the growth measurement for that test-tube.

	strain	shelf	growth
1	sensitive	middle	5.434952
2	sensitive	bottom	5.237206
3	sensitive	middle	5.028470
4	sensitive	bottom	5.321346
5	sensitive	bottom	4.730719
6	sensitive	top	10.625011
7	sensitive	top	10.914012
8	sensitive	bottom	4.879497
9	sensitive	middle	4.688804
10	sensitive	bottom	4.955090

5.3   Generating a simulated data set for our focal experiment

OK, now we have a clear idea of what our data frame containing our actual experimental data will look like, the next step is to get R to carry out an imaginary experiment and generate a data frame containing imaginary growth rate data. And as you already know from Chapter 4, to do this will require some decisions about effect sizes and variation. We also need to give some thought to how we communicate this information to R to allow it to generate our imaginary measures. In Chapter 4, we simply specified the means and standard deviations of our two treatment groups. We then told R to draw random numbers from normal distributions with the appropriate mean and standard deviations to generate our imaginary data. Here we are going to use the same general principle, but we will implement it in a slightly different way. The reason for this is that our new method will more naturally extend to other more complex designs.

So let's think again about our design and the information we will need in order to be able to simulate it. The design is summarized as a table in Fig. 5.2.

Our design has two factors, and so is referred to as a two-factor or two-way design. The treatment factor has two levels (*Sensitive* and *Resistant*), and the shelf factor has three (*top, middle* and *bottom*). It is fully factorial, which means we have all combinations of both factors, and so our table has six cells in it. In each cell of the table, we have included the information that R will need to simulate our experiment, which in this case is the mean and standard deviation for each combination of the factors. So where do these numbers come from? As in the previous chapter, they come from our knowledge of the biology of the system and the decisions we make about the size of effect we want to be able to detect with our study. So let's start with the biology.

Imagine that you have actually worked extensively with the *Sensitive* strain in previous studies and have good estimates of the average growth rate on all three shelves of your growth chamber. The average rate on the middle shelf is five divisions, whilst it is slightly higher on the top shelf (5.3 divisions) and lower on the bottom shelf (4.9 divisions). These same studies give an estimate of the standard deviation of 0.3 divisions for the mean growth rate on each shelf. OK, that's three of the six boxes filled in. Now we need to think about the *Resistant* strains growing on each shelf. Our study is aiming to see whether resistance comes with a cost in a herbicide-free environment (such as our growth chamber), so let's assume we want to be able to detect reductions in the *Resistant* strain growth rate of 0.2 divisions relative to the *Sensitive* (which

Fig. 5.2 Design table for our imaginary experiments, holding the mean and standard deviation of the growth rates (measured as the number of cell divisions in a 24-hour period) that we are assuming for our simulated studies. Notice that the alternative hypothesis allows for different growth rates between genotypes and between different shelves in our growth chamber.

Shelf	Strain	
	Sensitive	Resistant
Top	5.3 (0.3)	5.1 (0.3)
Middle	5 (0.3)	4.8 (0.3)
Bottom	4.9 (0.3)	4.7 (0.3)

amounts to about a 5 per cent reduction). Before we call it a day with our imaginary estimated values for the *Resistant* strain, we need to briefly consider if we anticipate any interplay between predictor factors that might affect the values expected when we test our null hypothesis.

5.3.1 Interactions

And so, here, a final issue we will need to consider is whether we expect the difference in growth rate between *Sensitive* and *Resistant* strains to be the same on the three growth shelves, or whether there is the potential for the cost of resistance to vary amongst the growth shelves. In statistical terms, do we expect there to be an interaction between the treatment and shelf factors, or do we expect them to affect growth rate independently? For the current study, we are going to assume that the factors act independently, and so to calculate the mean for the *Resistant* strain on each shelf we simply subtract 0.2 from the mean for the *Sensitive* strain on the same shelf. This gives us our *Resistant* values of 5.1 for the higher shelf, 4.8 for the middle shelf, and 4.7 for the bottom shelf, and we will imagine that the standard deviation is again 0.3 divisions.

For other studies, we might be interested in being able to detect interactions between factors that might not act independently, and in that case you would need to make a decision about how the effect of one factor varies across the levels of the other factor. So, it is worth spending a few minutes thinking about how we might do that with this study. Imagine that we expect costs of resistance to be lower under optimal growth conditions, and we think that the conditions on our middle shelf are optimal, whilst the top and bottom shelves are slightly less than optimal. In that case, we might assume that the *Resistant* strain grows 0.2 slower on the middle shelf, and 0.3 slower on the top and bottom shelves. If we use these differences to calculate the means of the *Resistant* strains, then this interaction can be incorporated into our simulated data. You can hopefully see how you could include all sorts of different interactions, if the biology of your system required them. We explore this issue in much more detail in Chapter 7.

5.3.2 Simulating data to put in our data frame

Figure 5.2 is our summary of the world that we need to simulate for our imaginary experiments. It contains all the information that R will need to allow it to generate and analyse an imaginary data frame. Let's start with the completely randomized design. Remember, this is where we are randomly allocating 30 tubes to each shelf in the growth chamber, ignoring the genotype. Script 5.1 is one way to do this (and if you don't want to type it in yourself, go and download it from the ESM, and then load the script into RStudio). Since this book is not supposed to be a programming manual, we are not going to go through this script line by line. However, we have included a large amount of annotation that tells you what each bit of code is doing. We would encourage you to play with the script and try to understand how the code is working.

```
#Script 5.1
#Script to generate data for a design where equal
#replicates of two treatments are randomly allocated to
#spaces across three shelves in a growth chamber with equal
#numbers of tubes on each shelf.
```

```
#Clear out R.
rm(list=ls())

#Matrix to store the means of the 6 combinations of
#factors, indexed as follows
#(sens/Shelf1, sens/Shelf2, sens/Shelf3),
#(res/shelf1, res/shelf2, res/Shelf3).
cell.means <-rbind(c(5.3,5,4.9), c(5.1, 4.8, 4.7))

#Set the value for the standard deviation.
sd <-0.3

#Set up a variable (reps)that contains the number of
#replicates per treatment. So total sample size is reps*2
#N.B.this Script will only work if this is a multiple of 3.
reps <- 45

#Set up a vector containing the number of
#1s, 2s and 3s for the design corresponding to the
#number of replicates on each shelf
shelflist <- rep(1:3, ((2*reps)/3))

#Shuffle the shelflist vector so that tubes will be
#assigned to shelf in a random order. NB For ease of
#reading we have split this line of code over two, but
#R will read it as a single line
randomshelf <- sample(shelflist, (reps*2),
            replace = FALSE, prob = NULL)

#Labels for the two levels of the treatment factor
treatmentnames <- c("sensitive", "resistant")

#Labels for the different levels of the shelf variables
shelfnames <- c("top","middle","bottom")

#Set up 3 empty vectors that we will use to store values
#in later. These will be used to generate our data frame.
strain<-c()
shelf<-c()
growth <-c()

#set up a counter to keep track of the number of reps
#created so far. This is used to randomly pick shelves.
count<-0

#The nested loops start here.
#t loops through the two treatments whilst r loops
#through the replicates for each treatment
  for(t in 1:2) {
    for(r in 1:reps) {
```

```
#Add 1 to the counter each time we start the loop
#To keep track of the current replicate's number.
count<-count+1

#Sets the shelf number for this rep to the value drawn
#from the random shelf allocation list created above.
s <- randomshelf[(count)]

#Generates the growth value for this rep and adds it to
#the vector (growth).
#Does so by drawing a random number from a normal
#distribution with the appropriate mean for the current
#factor combination (extracted from the cell.means table)
#and sd defined above.
growth <-append(growth, rnorm(1,cell.means[t,s],sd))

#Adds the label for the treatment and shelf for the
#current replicate to the appropriate vector.
strain <- append(strain, treatmentnames[t])
shelf <- append(shelf, shelfnames[s])

        }
    }
#Nested loops end here.

#now to combine vectors into a data frame
growth.data <-data.frame(strain, shelf, growth)

#Carry out out the analysis using a linear model
#store the result in an object called analysis.
analysis<- anova(lm(growth~shelf+strain))

#Display the data frame and analysis results.
print(growth.data)
analysis
```

Instead of focusing on coding, we will describe in general terms what the script is doing. The first thing the script does is set up a 2 by 3 table which contains the means of our six factor combinations. We have used the values from Fig. 5.2, but you can play around with the numbers to explore different effect sizes. The script also defines a variable, reps, that tells R how many replicates of each treatment we want and a variable, sd, that tells R the value for the standard deviation that we want to use for our random variation.

To generate all of our imaginary measurements, we use something called nested loops. We introduced the concept of a loop in the previous chapter as a way of getting R to repeat something a set number of times. Nested loops simply involve setting up a number of loops one inside the other. Whilst this is definitely not the most elegant way to do things (and R purists will be horrified), we think it is a straightforward and intuitive way to attack this problem. As you become a better R programmer, you will find other ways to do the same thing more efficiently, and you can find some tips on how to do this in the ESM associated with Chapter 4. But for now we will stick with loops. In this case, we will use two loops. The outer loop, which we will call the treatment loop, will repeat all of the actions inside it as many times as we have levels for our treatment. In this case we have two treatment levels, *Sensitive* and *Resistant*, so everything inside the treatment loop will be repeated twice, once for each treatment. The inner loop, which we will call the replicate loop, is completely enclosed in the outer loop. This loop is set to repeat as many times as we have replicates in each treatment. It is inside this loop that we put the code to generate each data point. To generate each data point, we also need to tell R which shelf it is on. Since we are going to randomly allocate tubes to shelves (with the constraint that we want exactly 30 tubes on each shelf), we need some way to include this random allocation in our experiment. Let's label the top shelf 'shelf 1', the middle one 'shelf 2', and the bottom one 'shelf 3'. The trick we use to do this is to get R to generate a list containing 30 1s, 30 2s, and 30 3s, and then shuffle this list to produce a random permutation. The first virtual test-tube that we generate is put on the shelf corresponding to the first number in the shuffled list, the second to the shelf corresponding to the second, and so on and so on until all 90 test-tubes have been allocated to shelves.

And now, we are finally ready to generate each data point using the code inside the replicate loop. For each pass through this loop, R generates a single data point by drawing a random number from a normal distribution with the mean for the appropriate treatment and shelf (taken from the table we generated at the top) and the standard deviation we have told R to use. This data point is added to a vector (a list) which will eventually contain all 90 measures. At the same time, a label for the appropriate treatment (i.e. *Sensitive* or *Resistant*) is added to a second vector and a label for the shelf (i.e. Top, Middle, or Bottom) to a third vector. Once we have generated all of our data points and our nested loops are finished, these three vectors are combined to produce the columns in our data frame. R helpfully also gives each column in the data frame the same title as the name of its vector. The nested loop process is shown diagrammatically in Fig. 5.3.

The final step is to analyse our simulated data and see whether we can detect the cost of resistance that we know is there to be detected. The analysis we have used here is a general linear model, which allows us to look at the effect of strain on growth rate of *C. reinhardtii* having controlled statistically for the overall shelf-to-shelf differences we know exist. This is most definitely not a statistics textbook, so we won't discuss why we chose this analysis. But if this really was your own study, you would have already given this matter a great deal of thought. Run the script and go and look at the command window. We have included a couple of print commands so that you can see both the imaginary data set that it produces and the output from the analysis, which will look something like Fig. 5.4. You will generate different random numbers, so your table should be in the same form but your calculated figures will be different.

The p-value we are interested in here is the one found in the second line of the analysis of variance table (the line that starts with strain). Is yours significant or not significant? Have you got the right or wrong answer (i.e. we know there really is a difference between genotypes; would your analysis of the sample cause you to conclude that)? Of course, to estimate power we will need to do this several thousand times, and we will give you a script to do that in a minute. However, when you are writing your own simulations, it is a good idea to stop at this point and make sure that everything is working as expected. Does the data frame look as you expect it to? Does it have the right number of columns,

Fig. 5.3 A diagram to illustrate how the nested loops can be used to generate the data for our imaginary experiment. In the example, the outer loop goes around twice: firstly for the *Sensitive* replicates and secondly for *Resistant* replicates. The inner loop goes around 45 times for each turn of the outer loop, and at each turn generates a measurement for that replicate as well as assigning it to a shelf.

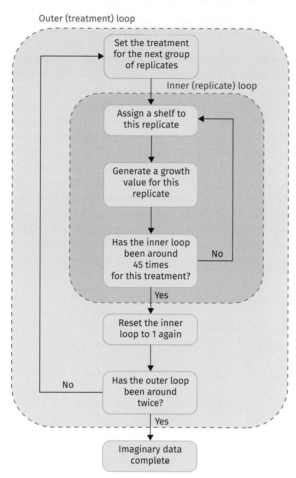

Fig. 5.4 Output from the GLM evaluation of our fully randomized design. Since this is based on the actual random samples that we drew for our imaginary experiment, don't worry if the values on your table differ (possibly substantially) from the actual values we have here.

```
Analysis of Variance Table

Response: growth
          Df  Sum Sq  Mean Sq  F value   Pr(>F)
shelf      2  1.7397  0.86985  10.860    6.231e-05 ***
strain     1  1.8140  1.81398  22.648    7.785e-06 ***
Residuals 86  6.8880  0.08009
—
Signif. codes:  0 '***' 0.001 '**' 0.01 '*' 0.05 '.' 0.1 ' ' 1
```

and contain the right number of data points in each treatment? If you have simply downloaded the script from our website, then we hope the answer to all of these questions is yes, but if you have typed it in by hand, then it is definitely worth checking.

5.3.3 Rinse and repeat: moving from a single simulation of your experiment to generating an estimate of power

Once we are satisfied that our imaginary experiments are working as expected, the final step is to automate the process so that R will run thousands of experiments one after the other, analyse each one for us and count the number that get the right answer. You already know how to do this from the previous chapter; we simply take the script for the individual experiment and put the whole thing in another big loop that will run for as many experiments as we want to carry out. We have done this in Script 5.2. We have also added some lines to extract the p-value that we are interested in for each experiment and store each one in a vector (which is what R calls a list of numbers). If you want more information on how we extract the right p-value, you can find that discussed in the ESM. Then once we finish the required number of runs, we simply ask R to count the number of p-values in our list that are significant (it has functions to do that). As with the previous script, we have included extensive commentary on the code so that you can understand how it works and what it is doing.

```
#Script 5.2
#This is the same script as 5.1 but embeded in a loop to
#allow multiple runs of the experiment. Detailed annotation
#of the main body can be found in Script 5.1
#only new items or changes are annotated here.


rm(list=ls())

cell.means <-rbind(c(5.3,5,4.9), c(5.1, 4.8, 4.7))
sd <-0.3
runs <- 100000
reps <- 45


#create an empty vector where we will store the p values.
plist <- c()

#big loop starts here.
for(experimentloop in 1:runs) {

shelflist <- rep(1:3, ((2*reps)/3))
randomshelf <- sample(shelflist, (reps*2),
          replace = FALSE, prob = NULL)
treatmentnames <- c("sensitive", "resistant")
shelfnames <- c("top","middle","bottom")

strain<-c()
shelf<-c()
growth <-c()
count<-0
```

```
  for(t in 1:2) {
    for(r in 1:reps) {

count<-count+1
s <- randomshelf[(count)]

growth <-append(growth, rnorm(1,cell.means[t,s],sd))
strain <- append(strain, treatmentnames[t])
shelf <- append(shelf, shelfnames[s])


      }
  }

growth.data <-data.frame(strain, shelf, growth)

analysis<- anova(lm(growth~shelf+strain))

#Extract the appropriate p value and store
#it as a variable called p
p <- analysis[2,5]

#Add this p value to our running list of p values
plist <- append(plist, p)

}

#Count the number of items in plist that are
#less than 0.05 (i.e. significant) and store this
#in a variable called significant.
significant <- length(which(plist<0.05))

#Calculate power by dividing the number of significant
#results by the total number of runs, and sends the
#result to the Console.
print(paste("power =", (significant/runs)))
```

Why not play around with the script to run some power analyses? Remember, you need to use your mouse to select the whole of the script and then click on run. You could start with the planned experiment with 45 tubes in each treatment group and see what power you would expect with this design. The program will take a few minutes to run (doing 100,000 experiments and analysing them takes time, even if they are imaginary!). On Nick's desktop it takes about 3 minutes, so don't despair if nothing happens for a while. Eventually the program will output the power to the Console. For example, as we are running it now, the line 'power = 0.86832' has appeared. Your value should be quite similar, since we have run a lot of imaginary experiments and should be getting a fairly precise estimate of the power. Now you can modify the script to try different experiments by changing the line:

```
reps <- 45
```

to reflect different sample sizes. Perhaps try 30 or 36 replicates: note, because of the way the simulation is set up, it will only work with numbers of replicates that are multiples of three so that the total sample size will be spread evenly

Fig. 5.5 The impact of increasing sample size on the power of the simulated experiment using a completely randomized design.

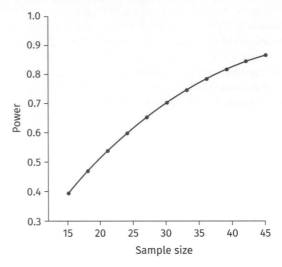

over the three shelves. We have tried this, and the results are shown in Fig. 5.5. We are not considering the possibility that we might use different numbers per shelf, and to do so would require a bit of script editing.

> ### Key point
>
> If you can produce simulated data organized in the same way as your actual experimental data, then conceptually (and practically) calculating power is just as easy for more complicated designs as it was for the simple designs of the previous chapter.

5.4   Comparing different designs

Part of the purpose of our simulations in this chapter is to allow us to use statistical power to compare different plans for allocating the individual tubes to the shelves in our growth chamber. As we emphasized in Chapter 2, some people see the purpose of power analysis as being to suggest how big your experiment should be, but we think power analysis is much more exciting than that. Power analysis can help you consider whether it is worth doing the experiment in the first place, how you might go about organizing your experiment, how big it should be, and how best you should collect and analyse your data. The more complicated the experiment, the more scope you have for adopting different designs. Since doing real experiments is often hard work, it makes sense to play around with imaginary experiments so as to make sure your real experiment is as good as it can be. Let's explore whether the design of the experiment we have been thinking of so far in this chapter is the most attractive.

The imaginary experiments we have just simulated used a completely randomized design, where the individual tubes were allocated at random to the positions in the growth chamber. This means we end up with the same number

(30) of tubes on each shelf, but there are likely to be different numbers of each of the two treatments on each shelf. We will now compare this with a randomized block design, where we ensure that each shelf has exactly 15 tubes of the *Sensitive* strain and 15 tubes of the *Resistant* strain, but still randomize the positions of the individual tubes across the shelf. Generally, statistical tests are at their most powerful when there are equal numbers of experimental units in each situation (for each combination of factors in our case). Hence, we might expect that there is a reasonable chance that the small amount of extra organization required to adopt a randomized block design might pay off in extra power. The question is, how much extra? We can easily turn our flexible simulation approach to addressing this question. Once you have written the code for one design, it's generally very easy to modify it to explore the power of another.

Have a look at Script 5.3. If you spent time getting to grips with the previous script, then hopefully much of this one should look fairly familiar to you. We have indicated where the changes are to help you compare the two. The only real difference between the two scripts is the way in which they generate the shelf information for each imaginary data point. Since we are not allocating tubes to shelves at random any longer, we replace that part of the script with another layer in our nested loops. We will call this our shelf loop, and it is wrapped around our treatment loop (which you probably remember is itself wrapped around our replicate loop). So our script will start by filling the top shelf with the 15 tubes of *Sensitive* algae, then move on to the 15 tubes of *Resistant* algae on the top shelf. It will then repeat this process for the middle shelf and finally the bottom shelf. It is also worth noting that the additional loop means we have to change what we mean by our replicate variable at the top of the script. In the previous script, it was the number of replicates of each treatment, whilst in this one, it is the number of replicates of each treatment on each shelf. Can you see why we made that change? The rest of the script is unchanged as it is doing exactly the same thing.

```
#Script 5.3
#Generates a balanced randomized block design where
#two treatments are allocated in equal numbers of replicates
#on each of 3 shelves in a growth chamber

#Clear out R
rm(list=ls())

#These lines specify the imaginary world.
cell.means <-rbind(c(5.3,5,4.9), c(5.1, 4.8, 4.7))
sd <-0.3
runs <- 100000

#Define the number of reps per treatment per shelf
#so that total sample size is reps*6. NB
#this is different from in Scripts 5.1 and 5.2.
reps <- 15

#Set up an empty vector to store our p values.
plist <- c()
```

```r
#The start of the loop for repeating the experiments.
for(experimentloop in 1:runs) {

#Labels for the two levels of the treatment factor
#and three levels of shelf
treatmentnames <- c("sensitive", "resistant")
shelfnames <- c("top","middle","bottom")

#3 empty vectors so store values in later.
strain<-c()
shelf<-c()
growth <-c()

#Nested loops to generate the data set
for(s in 1:3) {
  for(t in 1:2) {
    for(r in 1:reps) {

#Generate the growth value for this rep and add it to
#the vector (growth).see Script 5.1 for more details.
growth <-append(growth, rnorm(1,cell.means[t,s],sd))

#Add the label for the treatment and shelf for the
#current replicate to the appropriate vector.
strain <- append(strain, treatmentnames[t])
shelf <- append(shelf, shelfnames[s])


    }
  }
}

#Create the data frame.
growth.data <-data.frame(strain, shelf, growth)

#Analyse the data, extract the P value and add
#it to our P value vector.
analysis<- anova(lm(growth~shelf+strain))
p <- analysis[2,5]
plist <- append(plist, p)


}


#Determine the number of P values in our list
#that are significant.
significant <- length(which(plist<0.05))

#Output the estimated power to the Console.
print(paste("power =", (significant/runs)))
```

Fig. 5.6 The impact of increasing sample size on the power of the simulated experiments using a randomized block design compared with a completely randomized design.

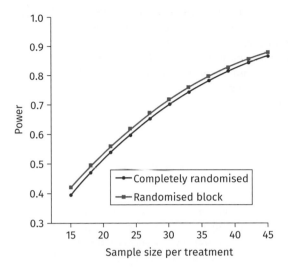

This script is actually more straightforward than the previous one. The removal of the random shelf allocation simplifies things a lot, so it is worth spending some time getting your head around the code. Once you are happy with it, try running it and estimating power for the originally planned 45 tubes per treatment group with the same effect size and shelf effects that we assumed before. Then you could run it again looking at the other sample sizes you examined for the previous design. What do we learn by doing this? The results from our runs are shown in Fig. 5.6.

So what have we learned from our simulations that can help guide our plans? Well the first thing to note is that, after all that, the different designs have very similar power over the range of sample sizes we have looked at. However, the randomized block design has more power than the other design at all sample sizes and, given the small amount of additional work in implementing this design with our study system, we would argue that it is worth it.

Probably the reason that randomized blocking doesn't buy us much more power is that relatively high sample sizes meant that we never strayed too far from a balanced design. We might not end up with 15 tubes of the two different types on our shelves every time, but it was probably pretty close to these values most times, say 14:16 or 13:17. If we were working with much smaller sample sizes, then the benefits of balance would have been stronger.

 Key point

Once you have the code to calculate power for one design of your experiment, it often requires very little extra effort to modify it to allow you to compare alternative designs.

5.5   Drop-outs: using power analysis to explore the possible impact of adverse events

We hope you are starting to see that one of the great advantages of simulating experiments to estimate their power is that rather than being limited to choosing from a number of off-the-shelf designs, we can tailor our simulations to exactly match the study we are planning to carry out. We will finish this chapter by thinking about the effects of another form of experimental complexity: experimental drop-outs.

In any real study, no matter how carefully we set things up at the start, there is a risk that things will happen such that we lose some of our replicates before the end. In crop trials, growth plots get destroyed by storms; in drug trials, some people may decide to leave before the end or fail to follow the procedures correctly. One of us even had a colleague whose co-worker accidentally drank part of their experiment (it's a long story!). From the perspective of this book, these losses will impact on the power of the study, but how much of an effect are they likely to have? Let's add drop-outs to our herbicide-resistance experiment and see. Since we have already decided that the randomized block design is the way to go, let's focus on the effects of drop-outs on this design.

As always, let's start by thinking about the biology of our system. Based on years of experience of working with the strain in this lab you know that, no matter how carefully you work, about one tube in ten will become contaminated and so not yield usable growth rate data. You have no reason to assume that the situation will be any different for the *Resistant* strain, so this translates into a probability of 0.1 that any tube in your study will become contaminated.

The way that we will incorporate this into our imaginary experiments is to get R to generate a uniform random number between 0 and 1 for each tube in an experiment. If that number is greater than 0.1, then our imaginary lab tech has done a good job and the tube has not become contaminated. In that case we generate a growth rate measurement for that tube in the same way as we did before and move on to the next one. However, what if the random number is 0.1 or less? This indicates that contaminants have gotten into that tube and we cannot use it in our data set. In that case, rather than generating a growth measure for our tube, we add the letters NA to our data frame. NA is the code that R uses to indicate that a value is missing (that is, Not Available or, as it is sometimes more poetically put, NA is an indicator of missingness!).

In the previous chapter, we introduced the idea of conditional operators in our script as a way of getting R to do different things depending on the outcome of some test. Specifically, we made use of what computer programmers call an *if* statement. This basically allowed us to tell R the following:

```
if some condition is met, do something
```

We used it to check if our p-value was significant (the condition), and if it was, to add 1 to our counter (the action).

Here, we want to do something slightly more complex, which is to get R to choose between two different actions based on whether some condition is met. That is, we want to check whether or not our random number is greater than the probability of contamination. If it is, then we want to generate a growth measure but, if it is not, we want to generate a missing value. To do so, we will make use of the if statement's big sister, the *if ... then ... else* statement. Again, put into rough English, this allows us to tell R to do the following:

```
if some condition is met, do action A, else do action B
```

if ... else statements are an extremely useful tool in producing scripts for imaginary experiments, and can be used to allow scripts to do different things in different circumstances. For example, in a more complex study, your planned analysis may proceed in different ways depending on an initial statistical test, and *if ... then ... else* would allow you to incorporate that flexibility into your script. To see the way we have coded it in the current example, have a look at Script 5.4. Find the new section (it is in the centre of the nested loops) and see if you can see how the code works.

```
#Script 5.4
#This is a modification of Script 5.3 to allow for the
#possibility of drop-outs due to contamination
#detailed annotation can be found on Script 5.3
#Annotation here is limited to new lines of code.

rm(list=ls())

cell.means <-rbind(c(5.3,5,4.9), c(5.1, 4.8, 4.7))
sd <-0.3
runs <- 100000

#Create a variable called dropout that contains
#the probability of contamination.
dropout <- 0.1

reps <- 15
plist <- c()

for(experimentloop in 1:runs) {

treatmentnames <- c("sensitive", "resistant")
shelfnames <- c("top","middle","bottom")
strain<- c()
shelf<-c()
growth <-c()

for(s in 1:3) {
  for(t in 1:2) {
    for(r in 1:reps) {

#Generates a random number by drawing from a uniform
#distribution between 0 and 1 to be used to test for
#contamination
u <- runif(1)

#A conditional statement to compare the random number u
#to the dropout rate.If it is greater than the probability
#of infection, generate a measurement normally. If not
#then add an NA (R's missing value indicator) instead.
#[note we have split the first line of code over two lines
```

```
#to make easier to read, but R reads it as a single line]
if(u > dropout) {growth <-append(growth,
      rnorm(1,cell.means[t,s],sd))
  } else {growth <- append(growth, NA)}


strain <- append(strain, treatmentnames[t])
shelf <- append(shelf, shelfnames[s])


    }
  }
}
growth.data <-data.frame(strain, shelf, growth)
analysis<- anova(lm(growth~shelf+strain))
p <- analysis[2,5]
plist <- append(plist, p)


}
significant <- length(which(plist<0.05))
print(paste("power =", (significant/runs)))
```

Now try running the script to see what effect the drop-outs have on your ability to reject the null hypothesis. The script is set to run the original planned study with 45 tubes per treatment, so why not start with that? Would your planned study still have enough power, even taking into account the potential for drop-outs? Or do you need to rethink your study? You can also very easily look at the effects of different drop-out rates by editing the value for the variable in the script; so go on, have a play. Figure 5.7 shows some of the results that we obtained. As you probably expected, the drop-outs reduce the power of the study, no matter what the original sample size. However, even with drop-outs at a rate of 10 per cent, our biggest study is still high powered. In contrast,

Fig. 5.7 The impact of increasing sample size on the power of the simulated experiments using randomized block design with no drop-outs compared with 10% and 20% drop-out rates.

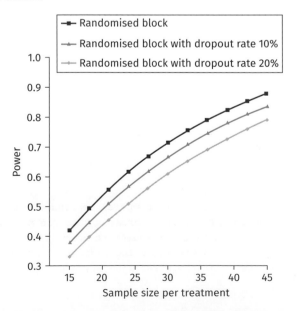

if we were considering reducing our sample size to 36 tubes per treatment, then the drop in power we see in our imaginary experiments at that level may cause us to question that decision. Ignoring drop-outs, we would expect our study to have a power of almost 0.8, but with drop-outs, this drops to about 75 per cent. This reduction in power occurs for two reasons. Firstly, we are reducing the number of data points in our analysis, so our effective sample size is smaller. Secondly, our design has probably become less balanced (the number of data points in each treatment group are unlikely to be the same).

5.6   Factors with more than two levels

The main factor in our study has two levels, *Sensitive* and *Resistant*. We kept things this way to keep them simple and allow you to focus on the important issue of how to set up your simulations. However in many studies, addressing the research question may require factors with more than two levels. For example, an experiment to examine the effects of tail-length manipulation on the attractiveness of male widow birds might include a group of birds with their tail length increased, a group with their tail length decreased, and a control group with the tail length unchanged. They might even include two control groups: one where the birds' tail is entirely unmanipulated, and a second procedural control where the tail is manipulated, but does not have its length changed. This adds additional complexities to the way we analyse the study, and consequently for our simulations. However, this chapter is already long enough, and you have already learnt plenty for one day. So for now, we just want to flag this issue. We will come back to it again in Chapter 7, where we will explore it in much more detail.

In this chapter, we have shown how the ideas developed in Chapter 4 can be extended to more complex study designs. We hope that you can also start to see how the scripts we have written can be modified reasonably easily to work for other similar designs that you might be considering, as might occur when adding additional levels for either of the factors (for example, if we wanted to add additional strains to our study). We leave it as an exercise for you to try to do this. Slightly more work would be to add an additional factor to our planned study. So suppose you wanted to assay your strains in three different growth media to see whether any cost of resistance depended on the richness of the growth medium, this would require an additional factor. We have also assumed that the number of replicates is the same for all combinations of our factors (at least before any drop-outs occur), which keeps things simple. However, it is only a relatively small step to change the script to allow the number of replicates to differ. We could even incorporate more complex biology, for example allowing the variation to be different for the two strains (although this change would also require a change to our method of analysis ... but that is another story). In the questions for you associated with this chapter, we challenge you to modify this study in a number of ways. Once you have got to grips with this material, we think you will be well on the way to being able to simulate almost any type of factorial design you might imagine (and such designs contribute a large proportion of those carried out).

 Key point

We can usefully employ power analysis to explore the effects of adverse events leading to unplanned loss of some of our data.

Summary points

- Sometimes additional factors in your experiment can arise for practical reasons rather than solely from the questions you want to explore.
- Imagine how the data from your experiment will look in a spreadsheet; your simulated data will have to be in exactly the same form.
- If you can produce simulated data organized in the same way as your actual experimental data, then conceptually (and practically) calculating power is just as easy for more complicated designs as it was for the simple designs of the previous chapter.
- Once you have the code to calculate power for one design of your experiment, it often requires very little extra effort to modify it to allow you to compare alternative designs.
- We can usefully use power analysis to explore the effects of adverse events leading to unplanned loss of some of our data.

Further reading

This seems a good point to mention some of our favourite books on R, and at the end of the next chapter we will name-check a few of our favourite statistics books.

Beckerman, A.P., Childs, D.Z. and Petchey, O.L., 2017. *Getting Started with R: An Introduction for Biologists*. Oxford University Press. [This book is a marvel for getting you started with R painlessly, and at the same time having a lot of wise things to say about how to do science. It's also a model of engaging (often funny) writing.]

Braun, W.J. and Murdoch, D.J., 2016. *A First Course in Statistical Programming with R*. Cambridge University Press. [This is a good book for discussing using R for the sort of simulations that we focus on.]

Zuur, A., Ieno, E.N. and Meesters, E., 2009. *A Beginner's Guide to R*. Springer Science & Business Media. [After several false starts, this is the book that converted Graeme to R.]

Exercises

5.1 Alter the code to add an extra strain called *super-resistant*. [This requires a change to the number of levels in the t loop (treatment), and the number of cells in the table of means.]

5.2 Modify the code to automatically explore a range of different sample sizes. Rather than have the replicate loop run for a specific number of times determined by the reps variable, you would replace the reps variable with a list where each value in the list is the sample size for that cell in the design. The rep loop will be set to the number of iterations required by pulling the value from the list (you can find an example of this type of script in the ESM associated with Chapter 4).

5.3 Modify the code to allow different sd for the strains. The single sd value should be replaced with a vector with sd values for the different levels of the factor. The call to the normal distribution will use the list element indexed by the current value of the looping variable.

5.4 Modify the code to consider two different drop-out rates for the two different strains.

6 EXTENSIONS TO OTHER DESIGNS

Learning objectives

In this chapter, you will learn to:

- Simulate studies where the predictor variable is continuous, and there is a straight-line relationship between the response and the predictor.
- Extend your simulations to include other kinds of relationships, such as quadratic.
- Deal with response variables that are not continuous, and variation that does not follow a normal distribution.

In Chapters 4 and 5 we showed you how R can be used to simulate several simple but commonly applicable experimental designs where we are interested in the effect of some factor with discrete levels on a continuous response variable; and where the random variation in that continuous variable can be assumed to follow a normal distribution. Even if you stopped your journey into power analyses there, you will already be able to simulate many of your planned experiments. However, we don't want you to stop there, because just a little extra exploration can open up a lot more flexibility for you. Thus, in this chapter, we are going to extend these ideas to another common type of biological study in which both the response and the predictor variable are continuous. For example, you might be interested in the relationship between water temperature and fish metabolism, and so you set up an experiment where you experimentally manipulate water temperature in aquaria along a continuous range and measure the oxygen consumption (and hence metabolism) of the fish in these aquaria. Or perhaps you are interested in whether the intestinal worm burden of school children in Africa affects their cognitive performance. To try to find out, you identify an appropriate sample of children, and for each you determine their worm burden and measure their performance in a cognitive test. You then look for a relationship between these two continuous variables. These are the kinds of study designs that lead to correlation or regression analyses (or one of their more complex relatives such as a logistic regression or analysis of covariance).

First we will simulate an experimental study where we have control over both the levels of the predictor variable and the number of replicates at each level. We will then remove the constraint of the previous chapter and allow for the possibility of having a different number of replicates at each level, and explore situations where the relationship is either linear or non-linear. Once we have that working, we will look at what happens if the levels of our predictor variable are not in our control. And, finally, we will explore ways to deal with situations where the property we want to predict is not on a continuous scale. An example of this would be if we looked to explore whether height is a predictor of whether a person gets married or not. Here we cannot expect to manipulate people's height; we must work with the natural variation that already exists, and each individual is either in one of two discrete states: married or not.

6.1   A simple linear regression

OK, let's suppose you want to know whether the amount of caffeine a student has had prior to tackling a logic problem affects the time it takes them to complete the problem. To find out, you are going to randomly allocate students to groups, and give each of them a drink with a controlled amount of caffeine. Half an hour later, you will ask each of them to complete a logic problem, and you will time how long this takes. You plan to have nine levels of caffeine in your experiment evenly spaced from 0 mg to 300 mg (which is equivalent to about four espressos!). Let's assume you will use the same number of replicates at each of our nine levels, and that you are interested in being able to detect whether there is a linear relationship between your response (time to complete the problem) and predictor variable (caffeine intake). Script 6.1 simulates just such an experiment, so type or load it into RStudio now and have a look. You will see there are many similarities in the basic structure of this simulation to the examples from the previous chapter. However, there are two main differences. The first is how we tell R about the design of our study. The way that we do this is to set up a table in R (technically in R speak, this is called a matrix) that includes one row containing the values for the levels of the predictor we will use in our study, and a second row containing the number of replicates for each level. The table for our starting design is shown in Fig. 6.1.

You will see that we have chosen values in the top row (the levels) that are evenly spaced, and values in the bottom row (number of replicates) that are all the same, but we could have put any numbers in these boxes to reflect different kinds of designs; so our simulation is going to be flexible.

We have then given this information to R by defining two vectors:

```
levels <- c(0,37.5,75,112.5,150,187.5,225,262.5,300)
reps    <- c(4,  4,  4,   4,     4,   4,     4,   4,     4)
```

The first contains the caffeine levels we are planning to use in our study, and the second stores the number of replicates at each level. And, finally, we use the following code,

```
design <- rbind(levels,reps)
```

to combine these two vectors into a single table (or matrix to use R terminology) and give the matrix a name (design) that we can refer to elsewhere in the script. We can also ask R to tell us what is in a particular cell in the table using the syntax:

```
design[row, column]
```

where row is the number of the row we are interested in and column the number of the column. So if we wanted to know the value of the third level of the predictor variable (i.e. 74 mg) in our design, we could use:

```
design[1, 3]
```

Fig. 6.1 Design table for the caffeine study. The first row contains the different concentrations of caffeine you are planning to use for each level of the continuous variable. The second row shows the number of replicate individuals at each level. In this case, we have assumed the same number at each level.

Caffeine concentration/mg	0	37.5	75	112.5	150	187.5	225	262.5	300
Replicates		4	4	4	4	4	4	4	4	4

whilst if we want the number of replicates at that level (i.e. 4), we type:

```
design[2, 3]
```

This will become very useful elsewhere in the program.

```
#Script 6.1
#Generates data for a continuous predictor with levels
#set by the researcher. Assumes a linear relationship
#between Y and X, with a defined slope and intercept.

#Clear out R.
rm(list=ls())

#Define the slope, intercept and sd of our imaginary world.
slope <- -0.1
intercept <- 230
sd <-30

#The next 3 lines define our design.
#Set up a vector containing the levels of our predictor.
levels <- c(0,37.5,75,112.5,150,187.5,225,262.5,300)
#Set up a vector containing the number of reps at each level.
#N.B. make sure these two vectors are of the same length.
reps <- c(4,4,4,4,4,4,4,4,4)
#Combine our vectors into a single matrix.
design <- rbind(levels,reps)

#We use the length function to automatically count the
#number of levels in our design,to be used in our loop below
nlevels <- length(levels)

#Set up empty vectors to store our x and y values
xval<-c()
yval <-c()

#The nested loops to generate the data set
#The outer loop goes through each level in turn
#The inner loop is set to the number of replicates at
#current level
  for(x in 1:nlevels) {
    for(r in 1:design[2,x]) {

#Generate a random deviation for the current data point
#by drawing from a normal distribution with mean of zero
#and sd as defined by user.
residual <-rnorm(1, 0, sd)
#Generate the predicted value for the current data point
#by substituting the current value of the predictor variable
#into the equation of the linear relationship.
predicted <- (intercept + (slope * design[1,x]))
#Generate the actual data point by adding the deviation
#to the predicted value, and add it to our y vector.
yval <-append(yval, (predicted + residual))
```

```
#Extract the current value of the predictor from the design
#matrix and add that value to the x vector.
xval <- append(xval, design[1,x])


        }
  }
#Nested loops end here.

#Combine our vectors into a data frame.
dataset <-data.frame(xval, yval)

#Analyse the data and store the result in an object
#called analysis.
analysis<- anova(lm(yval~xval, data= dataset))

#Display the data frame and results of the analysis.
print(dataset)
print(analysis)
#Produce a plot of the data.
plot(dataset)
```

The second difference is in the way we incorporate the biological effects into the simulation to allow us to calculate our imaginary data points. We are assuming that the relationship between our response and predictor variables can be represented by a straight line, and so we need to tell R what the slope and intercept of our straight line will be (since these two values completely specify a straight line). You can see we have done that at the top of the program by defining two variables, not very imaginatively called 'slope' and 'intercept'! Assume that we know from our previous work that a student who isn't especially caffeine-fuelled typically finishes this task in about 230 seconds, so we will set that as our intercept. We are expecting that caffeine would increase performance (i.e. reduce the time the task takes), and we would be interested in any strength of effect if it was more than 0.1 seconds improvement per mg of caffeine, so we will use that as our slope. As always, we also need to describe and estimate the variability in the system. As in previous examples, we are going to assume the random variation comes from a normal distribution and (let's assume), based on previous studies, we know that the sd for time to complete the task under controlled conditions is 30, so we will use that. This imaginary world that we will carry out our experiment in is shown in Fig. 6.2. As always, our question is: if this is what the world is like, how often will we be able to reject the null hypothesis of no relationship (or in mathematical terms, that the slope of the line does not differ significantly from 0)?

OK, now we have everything we need to generate our imaginary data. The general approach is going to be the same as in the previous chapter, and we make use of nested loops. In this case, our design only requires two. The outer loop is set for as many iterations as we have levels in our design (in this case nine), and rather than specifying this explicitly as in previous simulations, we let R work it out for us by using the line (near the top of the script):

```
nlevels <- length(levels)
```

The length function counts the number of items in our vector of the levels and stores this in a variable that we have called nlevels. We then use the variable

Fig. 6.2 An imaginary world in which caffeine improves the performance of individuals in the trial by 0.1 seconds per mg. The figures contain estimates of all the parameters we need in order to simulate the imaginary world. For any given level of caffeine we can generate a predicted time from the linear relationship. Thus, for 100 mg of caffeine, we predict a time of 220 s. We assume that the random variation around our predicted values is normally distributed, and so generate the actual measurement by drawing a random number from a normal distribution with mean 220 and sd of 30.

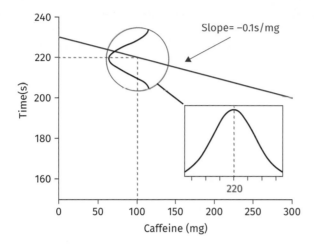

nlevels in setting up our outer loop. The advantage of doing things this way is that if we change the number of levels in our design, our loop is automatically set to match our design.

The inner loop then goes round as many times as there are replicates for the current level of the predictor. This is where we first make use of the design table, as each time we set up the inner loop for a new level of the predictor, we set the number of iterations of the loop using the number of replicates from the design table for that level. This means that if we had put different numbers in the second row of the design table, the inner loop would generate the appropriate number of data points for that level of the predictor.

For each turn of the inner loop, we generate an imaginary data point. To do this, we first generate a predicted value by grabbing the value for the current level of the predictor variable from the design table as we did for the number of replicates above, and plugging this into the equation of the line we specified with our slope and intercept variables. We then incorporate random variation by getting R to generate a random number drawn from a normal distribution with a mean of zero and an sd that corresponds to our estimate of the variation in our system, and we add this to our predicted value. After a few turns of the outer loop, and a few more of the inner loop, we have a dummy data set that matches both our design and our assumptions about the biology. Run the script to convince yourself. We have included lines to output the data frame and the analysis, as well as one to plot the data you have generated. You might want to play around with the script at this point. Try changing the number of replicates at some of the levels by editing the reps vector, and seeing if this works. You could even try adding additional values to the levels vector to increase the number of levels in the design (don't forget to also add the corresponding replicate values to the reps vector, or everything will go wrong and R will give you an error message). Don't worry if you break it, just close the script without saving and then reload the script and start again. Once you are happy with everything, it's time to actually calculate power. Load up Script 6.2 where we have put the current script into a big loop to allow it to run many times and use it to estimate the power of the current design. What did you get? We got 0.46921, which is not very promising.

```
#Script 6.2
#A version of Script 6.1 embedded in a loop to allow multiple
#runs of the experiment to be carried out automatically to
#estimate power. Annotation is only provided for aspects that
#are different to Script 6.1

rm(list=ls())

#Define our world.
slope <- -0.1
intercept <- 230
sd <-30

#The number of experiments we will use to estimate power.
runs <- 100000

#Define our design.
levels <- c(0,37.5,75,112.5,150,187.5,225,262.5,300)
reps <- c(4,4,4,4,4,4,4,4,4)
design <- rbind(levels,reps)
nlevels <- length(levels)

#Create an empty vector to store our p-values.
plist <- c()

#The start of the loop that repeats the whole experiment.
for(experimentloop in 1:runs) {

xval<-c()
yval <-c()

  for(x in 1:nlevels) {
        for(r in 1:design[2,x]) {

#Generate the data.
residual <-rnorm(1, 0, sd)
predicted <- (intercept + (slope * design[1,x]))
yval <-append(yval, (predicted + residual))
xval <- append(xval, design[1,x])

        }
    }

dataset <-data.frame(xval, yval)
analysis<- anova(lm(yval~xval, data= dataset))

#Extracts the appropriate p-value and stores it
#as a variable called p.
p <- analysis[1,5]
```

```
#Adds this p-value to our p-value vector.
plist <- append(plist, p)

#The big loop ends here.
}

#Counts the number of significant p-values.
significant <- length(which(plist<0.05))

##Calculates power as significant/runs and sends
#the result to the Console.
print(paste("power =", (significant/runs)))
```

Given our low power, let's play around with this script and see whether we can find a way to increase our power without involving more experimental subjects. We have assumed that we will have the same number of replicates at each of our nine levels of caffeine, but what if we change that? Let's keep the total number of replicates at 36, as in our default setup, but try some different allocation strategies by changing the numbers in the replicate vector at the top of the script. Note that, to keep things simple, we have written the script assuming the numbers in the replicate vector are non-zero and positive. If you put 0 for any level, the script will generate two replicates for reasons we won't bore you with! As your scripting skills improve, and especially if you want to have other people use your scripts, you will need to learn to put in various error checks to ensure people don't include a value that your code isn't designed to deal with. But for the time being, stick to positive integers. Alternatively, if you really want to remove a level from your design, remove it from both the replicate and levels vectors. Why not try having seven replicates at the lowest and highest levels, reducing the three middle levels by two to keep the total at 36? This increases our power to 0.59844; still not great, but a definite improvement. Let's try a design where we load our replicate towards the middle of the range with 6 replicates at the middle four levels, and two at the remaining levels. The power drops to 0.28, which is terrible and our subjects could be excused for feeling we were wasting their time if we used them in a study like that. Perhaps we can do better by changing the number of levels. Let's drop from nine levels to four, which allows us to have nine replicates at each level (but still cover the same range of caffeine values). This gets us up to 0.56. Let's throw all of our replicates at the two extremes: with 18 subjects at 0 mg and 18 given 300 mg. Our power has now jumped to a respectable 0.83. We do not think you will be able to do any better with 36 subjects (unless you increase the range of caffeine you are willing to use). This leaves us in a bit of an odd place, because we were planning a study with a continuous variable (level of caffeine consumed), but have ended up with a study with only two levels (either no caffeine or lots), which looks more like the factorial studies of the previous chapter. Our simulation has shown us that if we are interested in detecting a relationship between two variables and we are confident it is a straight-line relationship, then it is actually best to focus our replication at the extremes of our continuous range. However, with real studies this might be quite a big assumption to make and we might include some additional intermediate levels just to make sure our assumption is met. Our simulation can help us decide how we strike that balance.

💡 Key point

We can easily modify our approach to cope with continuous predictor variables as well as the discrete factors we have considered in the last two chapters.

6.2   Beyond the straight and narrow

So far we have assumed that the relationship between our response and our predictor variable can be represented by a straight line and so our description of our imaginary world was simply the intercept and slope of that line (and the variability around that simple general trend). However, in biology not everything follows a straight-line relationship. Fortunately, our script can easily be modified to simulate all sorts of other relationships. All we need to do is include variables for any coefficients that will be used to describe the relationship, and edit the line where we calculate the predicted value to use the appropriate equation. Going back to our previous example, what if we assumed that the relationship between caffeine intake and performance was actually as shown in Fig. 6.3? This is a curved (quadratic) relationship and can be represented mathematically by an equation of the form:

$$Y = C + m_1.X + m_2.X^2$$

where C, m_1 and m_2 are the three parameters that determine the actual shape of the relationship (note we have used an upper case C to avoid confusion in our

Fig. 6.3 An imaginary world in which the relationship between the speed of completing the task and caffeine intake is quadratic. The red line shows the assumed relationship between time and caffeine in our trial. The coefficients for the relationship shown are C = 230, m_1 = -0.1667 and m_2 = 0.00022, and the random variation around the predicted relationship is assumed to come from a normal distribution with an sd of 30. So for example, to generate a measurement for an individual that has consumed 100 mg of caffeine, we would generate a predicted time from the curve of 210 s, and then generate the actual measurement by sampling from a normal distribution with mean 210 and sd of 30.

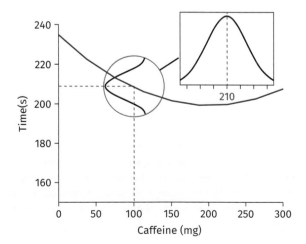

script, because R uses lower case c for something else). In the case of Fig. 6.3, C = 230, m_1 = -0.1667 and m_2 = 0.00022.

To alter our simulation to fit this, we would just need to replace the two variables in the current script, slope and intercept, with three variables (we have called them C, m_1 and m_2), and then alter the line where we calculate the predicted value to:

```
predicted<-(C + (m1 * design[1,x]) + m2 * (design[1,x]^2))
```

This is pretty easy, especially as we have done it for you in Scripts 6.3 (which is a single run of the experiment) and 6.4 (which repeats the experiment to estimate power). Of course, you also need to change the statistical analysis (we have done that for you too).

```
#Script 6.3
#Generates data for a continuous predictor with levels
#defined by the researcher. Assumes a quadratic relationship
#between Y and X, with coefficients m1, m2 and C defined by
#the researcher.

#Clear out R
rm(list=ls())

#These lines define our imaginary world.
#We have three coefficients to define the quadratic
#function: Y = C + m1*x + m2*x^2
C <- 230
m1 <- -0.1667
m2 <- 0.00022
sd <-30

#Sets up the design of our study.
levels <- c(0,37.5,75,112.5,150,187.5,225,262.5,300)
reps <- c(4,4,4,4,4,4,4,4,4)
design <- rbind(levels,reps)
nlevels <- length(levels)

#Creates two empty lists to store p-values for the
#linear and quadratic terms in the model.
plist1 <- c()
plist2 <- c()

#Empty vectors to store our response and predictor variable.
#Note, we have an extra vector which will allow us to have
#an additional column in our data frame for the X^2 values.
#This will make it easier to fit our model at the end.
xval<-c()
x2val<-c()
yval <-c()

#The nested loops start here.
  for(x in 1:nlevels) {
      for(r in 1:design[2,x]) {

#These lines generate the Y value for the current replicate
#using the quadratic relationship defined above and a
```

```
#random deviation drawn from a normal distribution.
residual <-rnorm(1, 0, sd)
predicted <- (C + (m1 * design[1,x]) + (m2 * (design[1,x]^2)))
yval <-append(yval, (predicted + residual))
xval <- append(xval, design[1,x])
x2val <- append(x2val, (design[1,x]^2))

    }
  }
#Nested loops end here.

#Combine our vectors into a data frame.
dataset <-data.frame(xval, yval, x2val)

#Carry out the analysis.
#We actually fit two models to the data.
#The first is the quadratic model (with results stored in
#analysis1). The second is the linear model (with
#results stored in analysis 2). This allows us to compare the
#power to discriminate the linear and quadratic relationships.
analysis1<- anova(lm(yval~ xval + x2val, data= dataset))
analysis2<- anova(lm(yval~ xval, data= dataset))

#Display the data and analysis results.
print(dataset)
print(analysis1)
print(analysis2)


##Script 6.4
#A version of Script 6.3 embedded in a loop to allow multiple
#runs of the experiment to be carried out automatically to
#estimate power. Annotation is only provided for aspects that
#are different to Script 6.3

rm(list=ls())

#Define our world.
C <- 230
m1 <- -0.1667
m2 <- 0.00022
sd <-30

#Defines the number of experiments used to estimate
#power.
runs <- 100000

#Our design.
levels <- c(0,37.5,75,112.5,150,187.5,225,262.5,300)
reps <- c(4,4,4,4,4,4,4,4,4)
design <- rbind(levels,reps)
nlevels <- length(levels)

#Create empty vectors to store the P values from the
#quadratic and linear analyses.
plist1 <- c()
plist2 <- c()
```

```r
#The outer loop to repeat the experiment starts here.
for(experimentloop in 1:runs) {

xval<-c()
x2val<-c()
yval <-c()

  for(x in 1:nlevels) {
    for(r in 1:design[2,x]) {

#Generate the current data point.
residual <-rnorm(1, 0, sd)
predicted <- (C + (m1 * design[1,x]) + (m2 * (design[1,x]^2)))
yval <-append(yval, (predicted + residual))
xval <- append(xval, design[1,x])
x2val <- append(x2val, (design[1,x]^2))

    }
  }

dataset <-data.frame(xval, yval, x2val)
analysis1<- anova(lm(yval~ xval + x2val, data= dataset))
analysis2<- anova(lm(yval~ xval, data= dataset))

#Extract our p-values.
p1 <- analysis1[2,5]
p2 <- analysis2[1,5]

#Add the p-values to the p-value vectors.
plist1 <- append(plist1, p1)
plist2 <- append(plist2, p2)

#The big loop ends here.
}

#Counts the number of significant P values in
#our p-value vectors and stores these in variables called
#significant1 (qudratic) and significant2 (linear).
significant1 <- length(which(plist1<0.05))
significant2 <- length(which(plist2<0.05))

print(paste("power to detect quadratic =", (significant1/runs)))
print(paste("power to detect straight line =", (significant2/
   runs)))
```

We have actually included two analyses, one with the quadratic term included and one without. This means we can look at the power of our design to detect any relationship at all, and then ask what the power is to detect the fact that it is actually quadratic. In this case, the answer is that our power to detect the fact that the relationship is actually curved is very low indeed. So whilst we can correctly reject the false null hypothesis of there being no relationship between speed in the trial and caffeine level 47% of the time, we can only reject the null hypothesis that the quadratic term is zero (i.e. the null hypothesis that the relationship is linear) in 6% of our runs. So, if power to detect non-linearity was our main motivation, our study would require some substantial rethinking. You could play around with the values and see whether you can get a design with more power, but detecting a subtle curve in a line like this is always going to take some serious replication.

Key point

Your experiment will generally have greater power to detect whether a relationship exists at all, than to detect features of the fine detail of that relationship.

6.3   When we don't control the values of our predictors

In all of our examples so far, we have been in control of the levels of our predictor variables. In the previous chapter, we chose the strains of *Chlamydomonas* to use, whilst in this chapter we chose the levels of caffeine to apply to each student. This will not always be the case; sometimes the value of a predictor variable for a particular replicate may not be in our control. This is most common in observational studies like the one we are about to consider, where we want to examine the relationship between two continuous variables; but since we don't actually manipulate the predictor, the variable's values are determined by the inherent variation in the random sample of subjects we pick for our study. However, this situation can also occur in more complex experimental studies. For example, in a study of the effects of a new blood pressure drug, we might want to measure and control for the body mass of our subjects, to allow us to include it in our analysis as a covariate. Or in an ecological study, we might repeat our experimental manipulation at several different field sites, and want to include this information in our design and analysis to account for any effect that this place-to-place variation has on our response variable. We will focus on the simplest case here, but the same approach can be extended to the other more complex situations.

Male hanging flies get their name as they hang from foliage offering a recently caught prey item to attract females. Here we are interested in whether the size of this nuptial gift offered by a male to a female during courtship affects how long he gets to copulate with her (which in turn is known to affect a male's paternity). To examine this, we are planning a study where we will randomly select 20 males from the local population and measure the size of their gift. We will then observe them until a mating takes place and measure how long the mating takes.

Our predictor variable (gift mass) is continuous (as in the previous example), with the difference that the value for any individual is not controlled by us. Instead we can think of the value as being drawn randomly from a statistical population (let's call this population all possible prey items that might be offered). So once again, we need to think about the biology. In this case, let's assume that the average mass of a nuptial gift is 4.5 mg, and that the variation in gift size follows a normal distribution with an sd of 0.3.

```
#Script 6.5
#Generates data for a study where the values of the predictor
#are drawn from a normal distribution with defined mean and sd.
#Y values are assumed to be linearly related to X, and the slope
#and intercept are specified by the researcher.

#Clear out R
rm(list=ls())
```

```
#Variables to define the imaginary world.
slope <- 14
intercept <- 0
sd <-2

#The parameters for the population we will
#sample our X values from
xmean <- 4.5
xsd <- 0.3

#Specify the total sample size of our study.
reps <- 20
#Set up empty vectors to store our x and y values.
xval<-c()
yval <-c()
#Start of a loop set to the number of replicates.
for(r in 1:reps) {

#Generate the current X value by drawing from a normal
#distribution, and add this to the vector of X values
thisx <- rnorm(1, xmean, xsd)
xval <- append(xval, thisx)
#Generate the predicted value for this replicate using
#the defined linear relationship.
predicted <- (intercept + (slope*thisx))
#Generate a random deviation for this individual
residual <- rnorm(1, 0, sd)
#Generate the y value by adding the deviation to the predicted
#value.
yval <- append(yval, (predicted + residual))

    }

#Combine our vectors into a data frame.
dataset <- data.frame(xval, yval)
    }

#Analyse the data and store the result in an object
#called analysis.
analysis <- anova(lm(yval~xval, data=dataset))

#display the data and the analysis.
print(dataset)
print(analysis)
#plot the data.
plot(dataset)
```

Now let's edit our previous script to model this experiment (you can download the edited version as Script 6.5). The first thing we can do is get rid of the design matrix, since we don't have any control over that in this study. We can replace this with a simple variable (we have called it reps again) that contains the total number of flies we will observe. We also need to include variables that tell R what the population of gifts will look like. We do this by giving R the mean gift size, and the associated standard deviation. And finally we need to include the parameters to describe the straight-line relationship between copulation time and gift size we are assuming is there to be detected (the intercept and slope again). In this case let's assume, based on other studies, that we are looking for a slope of 14 seconds per mg and an intercept of 0. As usual, the main change takes place in the inner loop where we actually generate the measurement for each individual. Here, since the value of the predictor is not determined by us, we need to get R to generate it for each individual by drawing from the normal distribution we have defined above. Then we simply use this value in the line of code that calculates the predicted value instead of pulling the value from a table. Other than that, everything else is much the same. Run the script a few times and check you understand how it is working. Then see whether you can embed it into a loop to estimate power. Would you carry out the study as described on the basis of its predicted power?

It is worth noting that often, in a study like the one just described, we are not interested in what the relationship between the variables is. Instead, our aim may be to simply determine whether or not they are associated (i.e. if the variables are correlated). In our case, this is asking if gift size seems to affect length of copulation at all. In principle, we can simulate such a study in a similar way. However, there is a slight twist. In the examples so far in this chapter, we have defined our imaginary world in terms of the mathematical relationship between the variables. Here our imaginary world is defined by the strength of the correlation between the two variables in the population that we are sampling from. Generating imaginary data based on a particular population correlation coefficient is a bit more tricky (and certainly a lot less intuitive) than using a specified relationship, and in practice, if we were going to do this we would actually make use of an R function called mvrnorm that is designed to do just that. We will not go into details here, but if you are interested, you can find more information in the Appendix to this book.

 Key point

In some studies, predictor variables will not be experimentally controlled; but our recommended technique can handle that without difficulty.

6.4   When things are not normal

So far we have assumed that the data we are collecting for our response variable are measured on a continuous scale, and that the random variation in our system is continuous and follows a normal distribution. This is true for many biological studies, and even when it is not, response variables can often be transformed to make it (at least approximately) true. However, there are situations where this will not be the case. For example, you might be working on disease transmission where your measurement is whether an individual is infected or not, which is clearly not a continuous variable. Or you might be counting the number of head lice on the heads of a sample of school children, which is not really continuous either, especially if the counts are relatively low (although if the average number is high it may behave like a continuous variable, as we discussed earlier for worm burden). Fortunately, it is relatively easy to extend what we have learnt so far to other kinds of data. So let's finish this chapter by looking at a situation where our predictor variable is continuous but our response variable is binary. Let's assume that we expose a number of individual beetles to differing doses of an entomopathogenic fungus (a fungal pathogen of insects), and we then record whether they become infected or not. How might we simulate this kind of data?

Let's focus on a particular beetle in our imaginary study. There will be some probability that it will become infected based on the dose of pathogen that it is exposed to. Let's assume that, for some dose that we are interested in, this is 0.2. Because of the way that R analyses this kind of data, we want our data set to include 1s for beetles that are infected and 0s for beetles that are not. How do we determine whether this beetle is 1 or 0? The answer is we use R's binomial random number generator which, in statistical parlance, will generate a random number of successes from a number of trials for a given probability of success. In this case (and counterintuitively from the beetle's perspective), success is going to be getting infected, the number of trials is 1 (we are only looking at one beetle at a time), and the probability is 0.2. The code to do this is:

```
rbinom(1, 1, 0.2)
```

where the first 1 is the number of random numbers we want, the second 1 is the number of trials, and the 0.2 is the probability of getting infected.

Of course that's fine if all beetles have the same probability of being infected, but in our example we are interested in modelling a situation where the probability increases with the dose of spores they get. So we need a way of determining the probability for this beetle, given the dose that it got. The way that we do that is to define mathematically the relationship between the probability of infection and dose. We will assume that this follows a logistic curve (because many dose-response curves do have this relationship). We won't worry about the mathematical detail here, all you need to know is that a logistic curve is an S-shaped curve, and its exact shape can be defined by two parameters. The logistic curve we will use here is shown in Fig. 6.4.

Fig. 6.4 An imaginary world where the probability of getting infected increases with the dose of spores. The red line shows the logistic curve that we have assumed relates infection probability to dose. Variation in infection status arises through a binomial process. Thus, to generate the infection status of an individual with a spore dose of 15 we use the logistic curve to predict the probability of infection, which in this case would be 0.29. We then use that probability as our binomial probability to determine whether the individual is actually infected or not.

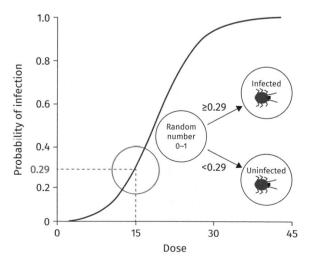

As we said, the logistic curve requires two parameters to describe it, so if we define those parameters as variables at the start of our program (we have called them m1 and m2 again), R can calculate the probability of infection for the dose level of the current beetle, in the same way as it generated a predicted value in the previous scripts. And then it can plug that probability into the rbinom statement to generate either a 1 or a 0. Then all we need to do is pop this at the heart of our nested loops and we can generate all the beetles we could ever desire. As always, we have provided a script to do this for you in the ESM, so go ahead and load Script 6.6 and have a play. Once again, we have given you a script to simulate a single experiment, show you the data and carry out the analysis. We leave it to you to embed this into a loop to repeat the process many times and so estimate power.

```
#Script 6.6
#Generate data for a design with a continuous predictor, but
#a binary response variable. The relationship between the
#probability of infection and the value of the predictor
#is assumed to be logistic, with coefficients defined by
#the researcher.

rm(list=ls())

#We set up two variables containing the model parameters
#for our logistic curve. m1 is the rate of turn, whilst
#m2 is the inflection point.
m1 <- 0.1
m2 <- 20

#Set up our design
levels <- c(5,10,15,20,25,30,35,40,45)
reps <- c(4,4,4,4,4,4,4,4,4)
design <- rbind(levels,reps)
nlevels <- length(levels)

#Set up empty vectors to store our x and y values.
xval<-c()
yval <-c()

#The loops to generate the dataset start here.
#The outer loop goes through the number of levels
#whilst the inner loop goes around the number of
#replicates of the current level.
  for(x in 1:nlevels) {
    for(r in 1:design[2,x]) {

#Generates the probability of infection of the current
#replicate based on the logistic curve.
predicted <- 1/(1 + exp(-m1*(design[1,x] - m2)))
#Generates a y value by drawing a number from a binomial
#distribution. This will be a value of either 1
#or 0, where 1 is infected, 0 is uninfected.
yval <-append(yval, (rbinom(1,1,predicted)))
```

```
#Adds the value for the predictor of the current replicate
#to the X vector.
xval <- append(xval, design[1,x])

    }
  }
#Nested loops end here.

#Combines our vectors into a data frame.
dataset <-data.frame(xval, yval)

#Carry out the analysis.Note we have split the line
#of code over two lines to make it easier to read.
#R will read it as a single line.
analysis<- anova(glm(yval~xval,
  family=binomial(link='logit'),data= dataset), test -
'Chisq')

#Display the data and the results of the analysis.
print(dataset)
print(analysis)
#Plot the data
plot(dataset)
```

> ### 💡 Key point
>
> Our simulation approach can handle discrete response variables, as well as the continuous ones we have considered up to this point.

6.5 Other distributions

R has many functions that can be used to generate data drawn from different distributions in the same way as we did for our individual beetles (and we discuss some of these in the Appendix). For example, if our beetle experiment had been done differently with the spores applied in the same way to replicate pots containing ten beetles under conditions where each individual has a 20% probability of infection, then we could have used:

```
rbinom(1, 10, 0.2)
```

to generate the number of infected individuals in the pot, assuming that the number is drawn from a binomial distribution with ten trials and a probability of success of 0.2 in each.

Going back to our head lice, we might want to model the chance variation in numbers per child as being drawn from a Poisson distribution. In this case, we need to decide the mean number of lice we expect, let's say it is eight. Then we can use:

```
rpois(1, 8)
```

to generate random numbers drawn from a Poisson distribution.

We could go on, but we won't. There are lots of other distributions that R can work with, and we will leave it to you to find out more if you need to. However, we hope this gives you a glimpse of what is possible and the confidence to find out more if you ever need to.

A book like this can never hope to cover every possible type of study that any biologist might like to carry out. Our aim in this chapter has been to extend the approaches you have learnt already to several other popular types of study. We think that small modifications of the scripts we have presented, and a little bit of thought, will allow you to estimate power for the vast majority of studies that biologists do. More importantly, now that you understand the general approaches to simulating experiments, and have a few tricks up your sleeve (and we hope a little bit more confidence than at the start of the book), we don't think you will find it that hard to come up with ways to simulate all sorts of designs.

 Key point

There are lots of different statistical distributions that you might want to use instead of a normal distribution; with a bit of searching you will find one that provides a good description of the biology of your situation.

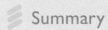 Summary

- We can easily simulate continuously varying predictor variables.
- An experiment that is powerful enough to give you a good chance of detecting that some kind of relationship exists might have substantially lower power to differentiate between similar descriptions of the nature of that relationship.
- There is no real technical challenge to calculating power by simulation if predictor variables are measured but are not under your control (such that the variation occurs naturally rather than by experimental manipulation).
- Similarly, extension to discrete response variables is easy too.
- And different descriptions of variability beyond the normal distribution can easily be accommodated too.

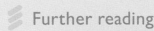 Further reading

As promised in the previous chapter, here are some of our favourite statistics books. But really you need to browse a few and find those that resonate with you.

Grafen, A. and Hails, R., 2002. *Modern Statistics for the Life Sciences.* Oxford University Press. [Every chapter has really intuitive nuggets of wisdom you won't find anywhere else.]

Holmes, S. and Huber, W., 2018. *Modern Statistics for Modern Biology.* Cambridge University Press. [In a slim volume this covers a huge diversity of topics—without being superficial—more a book to stimulate your mind than a technical manual, but has lots of handy R code.]

Quinn, G.P. and Keough, M.J., 2002. *Experimental Design and Data Analysis for Biologists.* Cambridge University Press. [There is a reason this has been cited over 10,000 times: it walks the line between being comprehensive yet well-integrated and thoughtful perfectly.]

Discussion questions

6.1   Discuss why we consider a child's parasitic worm burden to be continuously distributed.

6.2   Discuss why in the caffeine experiment our power was highest when we split all our replicates evenly between the two extreme levels.

6.3   Based on your knowledge of human biology, would you expect the relationship between caffeine intake and speed of solution to a logic problem to depart from linearity? If so, how?

6.4   Why might we expect a male hanging fly's copulation duration to be positively associated with size of nuptial gift?

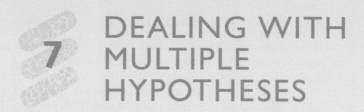

7 DEALING WITH MULTIPLE HYPOTHESES

Learning objectives

By the end of the chapter, you should understand:

- How you can evaluate the powers of different designs for each of several hypotheses separately.
- That no single design will likely offer exactly the same power for all these hypotheses.
- How ranking these hypotheses in terms of their importance to your study can help you find a design that offers suffi-cient power for all the hypotheses of prime importance.
- How having settled on such a design, you can evaluate whether the power available for testing ancillary hypothe-ses is sufficient to continue to include them in your study.

By this point in the book you should have a reasonable understanding of how you would go about estimating power for almost any kind of study that you might imagine using simulated data. You will also have started to develop a toolkit to allow you to implement these simulations yourself. You will know that one of the most important steps in this whole process is thinking about the hypothesis that you want to test.

To this point we have made life easy for ourselves by imagining that the study whose power you want to explore involves answering a single research question, and testing a single hypothesis. In reality, things are often more complex. You may be planning a study where you will explicitly measure multiple response variables to test a number of hypotheses. And even in a study with a single response vari-able, unless the design is extremely simple, you may be implicitly testing multiple hypothesis at once. In this chapter, we explore how you might take these issues into account when thinking about the power of your study. In exploring these issues, we think the real value of our approach to estimating power fully comes to light.

7.1   Several research questions in a single study

So far in the book we have essentially assumed that the study whose power you want to explore involves answering only a single research question. In reality you will often want to test several hypotheses in a single study. The bad news here is that it seems wise to carry out the type of power analyses that we have discussed for each hypothesis in turn. The good news is that in most situations, having written the code to perform power analyses for one hypothesis, only trivial modification will be required to carry out similar analyses for another hypothesis. Having carried out these power analyses for each of your hypotheses in isolation, you can then explore how to select a single experiment that offers you good power to address all your hypotheses. This exploration might involve further simulations. This is where you will likely find that no design will allow you to achieve the same target power (e.g. 80%) for all hypotheses simultaneously. This in turn might require some level of compromise. It might be useful to ask yourself for each hypothesis in turn how much dropping that hypothesis from your list of hypotheses that you initially felt you were interested in would allow reduction in the size of your experiment. It will often be that some hypotheses are more important to you than others, and if you are honest some of the hypotheses on your list have been added not because you are particularly interested in them, but because testing them seemed like a free bonus obtained through combining information that has to be collected to address the hypotheses you are really interested in. Your power analysis might quite often flag up situations where such bonus investigations are actually of little value, because an experiment optimized to address your questions of primary interest means you have very low power to address these bonus hypotheses. Increasing the size of your experiment to solve this might not be justifiable if these hypotheses really were not the prime drivers for the study.

We don't really have any particular tricks to offer you in this section. If your study is aimed at exploring multiple research questions, it doesn't seem unreasonable that you should have to work a little harder in your power analysis than for a simpler study.

 Key point

Increasing the number of research questions in your study will complicate your power analyses; but all the principles introduced for simpler studies still apply, and often code written to address one research question will only need slight modification to address another.

7.2   Designs that implicitly test multiple hypotheses

In a study with multiple research questions, it is obvious that you will be testing multiple hypotheses with separate statistical analyses. However, even in an apparently simple study with a single response variable, your planned analysis may actually test multiple hypotheses all at the same time. In these situations it is important to think carefully about exactly which hypothesis or hypotheses you really want to test, so you can calculate power accordingly.

In the remainder of this chapter, we will explore the two most obvious situations where this can arise: analyses that simultaneously test multiple hypotheses at once, and analyses that test a single hypothesis which is not actually the hypothesis you are really interested in. The latter in particular is an area where estimating power by simulation becomes particularly effective.

7.2.1 Main effect or interaction?

Let's return to the question we were exploring in Chapter 5 about the costs of herbicide resistance. You may recall that we explored this question by comparing the growth rates of *Chlamydomonas* in an environment that did not contain herbicide, looking to see whether resistant genotypes grew more slowly in such an environment (which would indicate a potential cost to herbicide-resistance). Let's assume that our study was successful, and we showed that resistance did indeed come at a cost exhibited in the absence of herbicide. We now want to further explore these costs. Imagine further that the reduction in growth that we detected was relatively small, but the experiment was carried out in conditions that are optimal for *Chlamydomonas* growth. Perhaps the costs would be greater in more stressful environments, so we are now going to design a study to test this. To do so, we will make use of the two-factor design shown in Fig. 7.1. As before, we will compare the growth of populations of resistant and sensitive *Chlamydomonas* in herbicide-free environments, but we will do this in both our original optimal environment and a second environment that is known to be more stressful to *Chlamydomonas*. Let's think about how we might estimate power for this design.

Firstly, let's consider the biology. Let's fill in the cells in our design table (Fig. 7.1) to reflect the imaginary world we want to use to estimate power. We will assume the growth rates of the sensitive and resistant strains in the optimal environment will be the same as we measured in our previous study. So the sensitive strain grows at a rate of 5 whilst the resistant one grows at a rate of 4.8, corresponding to a difference in growth rate of 0.2. These values have been added to the first column in Fig. 7.1. We also have a good idea of how the stressful environment affects growth rate of the sensitive strain, reducing its growth rate from 5 to 4.6. So the only additional thing we need to think about is: if the cost of resistance really does increase in the stressful environment, what is the smallest increase we would be interested in being able to detect? Let's assume that if the reduction in growth in the stressful environment increased from 0.2 to 0.4, that would be interesting to us, allowing us to fill in the

Fig. 7.1 A factorial study to examine whether costs of resistance are greater in a stressful environment. *Chlamydomonas* populations of either Sensitive or Resistant strains of the algae will be grown in either an environment in which they are known to grow well or in a stressful environment, where their growth rate is reduced. The values in the cells are the mean growth rates (measured as the number of cell divisions in 24 hours) and standard deviation that we are assuming in our imaginary world. Since the reduction in growth rate is greater in the stressful environment (0.4) than in the optimal environment (0.2) we are predicting a statistical interaction between the effects of our factors.

Factor 1 (strain)	Factor 2 (environment)	
	Optimal	Stressful
Sensitive	5 (0.3)	4.6 (0.3)
Resistant	4.8 (0.3)	4.2 (0.3)

final cell in our table. We will also assume that the standard deviation in growth rate is 0.3, the same as we measured in our previous study. Based on these assumptions, our imaginary world is shown in Fig. 7.1.

The final thing we need to think about is how we will analyse this data. In this case we would go for a two-way anova, and this is where we need to do a bit of additional thinking. If you are familiar with a two-way anova you will know that this single test actually tests three separate hypotheses at once. If you look at Fig. 7.2, you can see the kind of results we might get from an analysis like this.

The top two lines in the anova table (starting with strain and environment) are tests of what statisticians call the main effects, whilst the third row (starting with strain:environment) is a test of the interaction between the effects of the two factors. Which are we interested in? Let's think about each in turn. The main effect of strain is testing whether the herbicide-resistance causes an overall drop in fitness regardless of specific (herbicide-free) environment. We are confident that this is the case from our previous study, so this is not the p-value we are primarily interested in here. The main effect of environment is asking the question, do differences between our two environments affect growth rate (regardless of the particular strain)? Again, we have picked the stressful environment precisely because we are confident that it does reduce growth rate, so this is not the p-value we are looking for. The interaction asks the question, does the effect of one of the factors depend on the level of the other? Translating this into our study we might express this as, does the effect of herbicide resistance on growth rate depend on the environment? Hopefully, you can see that this is the question we are focused on in our study. And because of this, we should be estimating the power based on the significance of the interaction.

The reason that this matters is that the power of a study to detect an interaction is likely to be very different from the power to detect a main effect. You can see this in Fig. 7.3 where we present estimates of power for both main effects and the interaction for studies with a range of replicates per factor combination. If we have planned our study focusing on the main effect of environment we might decide on a design that only offers power to detect an interaction that is concerningly low.

Fig. 7.2 The results of an analysis of variance carried out on a study with the same design that we are planning. The top line of the table provides a significance test for the main effect of strain, the second line a test for the main effect of environment. The final column (Pr(>F)) gives the p-values for each. In this case both are highly significant. The third row provides a significance test of the interaction between strain and environment. In the case of this analysis, it has a p-value of 0.46 and is not significant.

```
                          Analysis of Variance Table
Response:           growth
                    Df      Sum Sq      Mean Sq     F value     Pr(>F)
strain              1       2.8092      2.8092      48.3834     7.045e-08  ***
environment         1       9.8358      9.8358      169.4044    2.503e-14  ***
strain:environment  1       0.0325      0.0325      0.5592      0.46
Residuals           32      1.8579      0.0581
---
Signif. codes:      0 '***'  0.001 '**'  0.01 '*'   0.05 '.'    0.1 ' '    1
```

Fig. 7.3 The estimated power for the two main effects and the interaction for our planned two-factor study with different numbers of replicates per factor combination. Despite the power for both of the two main effects being high for sample sizes of over eight, the power to detect the interaction is low across the whole range of sample sizes examined. The script used to generate these values can be found in the ESM for this chapter.

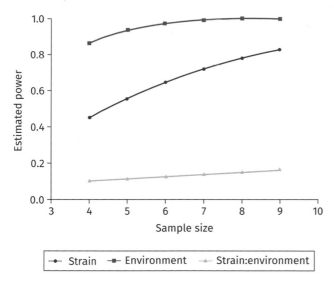

 Key point

If your analysis includes multiple potential hypothesis tests in the same analysis, the power of your planned study may vary dramatically for the different tests, so it is critical to think about exactly which test is central to your biological question.

7.2.2 Exploring specific comparisons in studies with multi-level factors

The second situation we will focus on in this chapter is one where you are designing a single-factor study, but your factor of interest has more than two levels.

So suppose you have designed a novel means of applying pesticide to wheat that you think will improve the performance of the pesticide compared to the current means of application, and consequently the amount of wheat that you get. You plan to randomly allocate an equal number of experimental plots to one of three treatments: 'no pesticide' (which will not have any pesticide applied), 'traditional' (which applies pesticide in the way currently used) and 'novel' (were the pesticide is applied in your new way). You will measure the yield of each plot at harvest. You have thought a lot about the imaginary world you want to simulate. The average yield of an experimental plot of wheat with no pesticide is 45 kg with a standard deviation of 7 kg, and average yield is increased to 52 kg when you apply pesticide in the traditional way. The novel

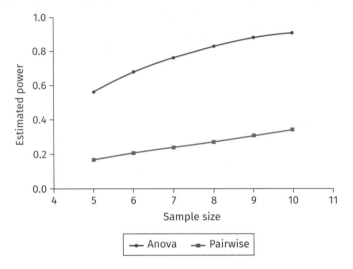

Fig. 7.4 Estimates of power for two different planned analyses of the pesticide study with different numbers of plots per treatment group. The red points indicate power estimates based on the Omnibus p-value from the anova, whilst the blue points represent power for a planned pairwise comparison of the traditional and novel treatment groups. For any given sample size, the power is lower for the pairwise analysis. The script used to generate these estimates can be found in the ESM for this chapter.

method of application is more time-consuming, but if it could increase yield under your test conditions to an average of 57 kg, it would be worth developing further. We assume that the standard deviations will be the same in all three cases. You decide that the way to analyse this data is to use a one-way anova, and so you write a script to do this for you. We have already done this, and the results of our efforts can be seen in Fig. 7.4 where the red line shows the estimated power based on this analysis. Looking at the figure, eight plots per treatment would seem to give you more than enough power, and so you decide to proceed with this number.

But there is a problem. The power analysis is based on how often your anova results allowed you to reject the null hypothesis, so let's think a bit more about exactly what null hypothesis this is testing. The null hypothesis of a one-way anova is that the samples in the three different treatment groups were all drawn from a population with the same mean. So, if we reject that null hypothesis, then the alternative hypothesis that we are actually accepting is that at least one of the three samples being compared comes from a population with a different mean from the other samples. This is not the same as saying that all of the samples come from populations with different means.

In the case of our study, we are expecting there to be a difference between the control and the traditional groups (we know that the pesticide works), and this alone would cause us to reject the null hypothesis, whether or not our new treatment performs any differently than the traditional treatment. So the power analysis shown by the red points is showing our power to answer a question that we are not really interested in. Put more generally, with studies that have multi-level factors, rejection of the null hypothesis of no difference amongst factor levels (the so-called **omnibus hypothesis**) is not usually the end point of the analysis. Instead, additional analyses would be carried out to test one or more specific hypotheses. In our case, we would follow up a significant result from our anova with a test of our specific hypothesis of interest (i.e. does the new treatment perform differently compared to the traditional treatment?).

There are literally hundreds of textbooks written on ways you might do this, and there is considerable discussion in statistical circles about the best approach to take. This is not the place to enter into such debates. So instead we will simply demonstrate one approach that might be used. If you or your statistical consultant prefer an alternative method, we don't think you will find it hard to implement that instead.

In this case, as a follow-up to a significant result from our initial anova we would get R to generate a single pairwise comparison between our two pesticide treatment groups. If we incorporate this additional step into our analysis, and use the p-value from the pairwise comparison as our basis of our power estimate, then our power is now focused on our ability to answer the question we are actually most interested in (not the more general question addressed by the main anova). We hope you will agree that this seems the sensible thing to do. And the blue line in Fig. 7.4 shows the results of our labours. As you might intuitively expect, our power has dropped compared to the results based on the p-value from the anova, as we are looking for a much smaller difference than the difference between our control and our two treatments. However, our simulations allow us to put a number to our intuition, and plan the right experiment for our actual hypothesis of primary interest.

Key point

If any of the factors in your study have more than two levels, make sure you have thought about exactly what specific comparisons you are actually interested in, and base your power analysis on tests of those specific comparisons.

7.3   Conclusion

The issues considered in this chapter, we think, provide one of the clearest illustrations of the value of our approach to power analysis through simulations. When you determine power based on simulated data sets, it requires you to use the exact approach to the analysis of your imaginary data that you will use for your real data. Put another way, if you can design a procedure for R to analyse your data (and given the power and flexibility of R, you probably can), no matter how bespoke your procedure is, simulation will allow you to determine power based on that actual analysis (not on some other analysis that is a bit like the one you had in mind but not exactly, but at least there is a package online that has it as one of its defaults). In our example, we followed our anova with a single pairwise comparison. In another study, your hypothesis might involve being able to detect significant differences between three different treatments and a single control group. In that case, you could design an analysis that examined those three pairwise comparisons, and base your power on the number of imaginary experiments that correctly generate significant results for all three comparisons. Or perhaps you are planning with a control group, a second procedural control group and a third experimental group. In that case, a positive result would require that the two types of control are not significantly different, but that the experimental group differs from them both. Again, you could design a series of planned comparisons to test this specific hypothesis, and use it to analyse your simulated data. So, basically, if you can imagine an analysis, and implement it in R, then you can use what you have learned to estimate the power of your study. In the next chapter we will expand this idea further and think about how we can extend our approach to look at statistical approaches other than null-hypothesis testing.

```
#Script 7.1
#A two-factor design to estimate the power of both of
#the main effects and the interaction.

#Matrix of means for the 4 combinations of genotype
#and environment indexed as follows
#(sens/optimal, sens/stress), c(res/optimal, res/stress).
cell.means <-rbind(c(5,4.6), c(4.8, 4.2))

#The value for the standard deviation.
sd <-0.3

#The number of imaginary experiments.
runs <- 100000

#The number of replicates per combination of genotype
#and environment.
reps <- 9

#Three empty vectors to store the p-values in.
strainplist <- c()
environmentplist <-c()
interactionplist <-c()

#Big loop starts here.
for(experimentloop in 1:runs) {

#Labels to be used for the genotypes and environments.
treatmentnames <- c("sensitive", "resistant")
environmentnames <- c("optimal","stressful")

#Empty vectors to store the columns of our data frame.
strain<-c()
environment<-c()
growth <-c()

#Nested loops to generate the data.
for(s in 1:2) {
   for(t in 1:2) {
      for(r in 1:reps) {

#Generates the growth value for this rep and adds it to
#the vector (growth).

growth <- append(growth, rnorm(1,cell.means[t,s],sd))

#Adds the label for the strain and environment for
#this replicate to the appropriate vectors.
strain <- append(strain, treatmentnames[t])
```

```
environment <- append(environment, environmentnames[s])

    }
  }
}

#Combine our three vectors into a data frame.
growth.data <- data.frame(strain, environment, growth)

#Analyse the data, and save in an Object called analysis.
analysis <- anova(lm(growth~strain*environment))

#Extract the appropriate p-values.
strainp <- analysis[1,5]
environmentp <- analysis[2,5]
interactionp <- analysis[3,5]

#Adds the current p-values to our p-value vectors.
strainplist <- append(strainplist, strainp)
environmentplist <- append(environmentplist, environmentp)
interactionplist <- append(interactionplist, interactionp)

#The big loop ends here
}

#Counts the number of significant p-values in each vector.
significantstrain <- length(which(strainplist<0.05))
significantenvironment<-length(which(environmentplist<0.05))
significantinteraction<-length(which(interactionplist<0.05))

#Outputs our power estimates to the Console.
print(paste("power:strain =", (significantstrain/runs)))
print(paste("power:environment =", (significantenvironment/
   runs)))
print(paste("power:interaction =", (significantinteraction/
   runs)))


#Script 7.2
#A one-factor design where the factor has 3 levels
#Analysis based on Anova followed by a planned pairwise
#comparison. Power estimated for the Omnibus test
#and the comparison of interest.

#The means and sd of the three levels of our factor.
treatment.means <-c(45, 52, 57)
sd <- 7

#The number of imaginary experiments that we want.
runs <- 10000

#The number of replicates per level of the treatment.
reps <- 10

#Empty vectors to store the p-values for the Omnibus test
#and planned comparison.
omnibusplist <- c()
plist <- c()
```

```
#Experiment loop starts here.
for(experimentloop in 1:runs) {

#Labels for the levels of our treatment.
treatmentnames <- c("control", "traditional", "novel")

#Vectors to store the columns in our data frame.
treatment <-c()
yield <-c()

#The nested loops start here.
for(t in 1:3) {

      for(r in 1:reps) {
#Generates the yield value for this rep and adds it to
#the appropriate vector (yield).

yield <-append(yield, rnorm(1,treatment.means[t],sd))

#Adds the label for the treatment for this replicate
#to the appropriate vector (treatment).
treatment <- append(treatment, treatmentnames[[t]])

    }
  }

#Nested loops end here.

#Combine our vectors into a data frame.
yield.data <-data.frame(treatment, yield)


#Carry out the analysis. We use
#a linear model to fit our anova and generate the
#Omnibus p-value.
analysis <- anova(lm(yield~treatment))

#Extracts the Omnibus p-value and stores it as a variable
#called p, and also adds this to our list of Omnibus
#p-values.
p <- analysis[1,5]
omnibusplist <- append(omnibusplist, p)

#If the omnibus result is significant, we do a pairwise
#comparison between traditional and novel and replace our
#Omnibus p-value with the result of the comparison.
if(p < 0.05){
comparison <- pairwise.t.test(yield, treatment, p.adj = "none")
p <- comparison$p.value[4]
}

#Adds the current p-value to our list of pairwise
#comparison p-values.
plist <- append(plist, p)

#The big loop ends here.
}
```

```
#Counts the number of significant p-values for each test.
significant <- length(which(plist<0.05))
omnibussignificant <- length(which(omnibusplist<0.05))

#Outputs the estimated powers to the Console.
print(paste("power =", (significant/runs)))
print(paste("omnibuspower =", (omnibussignificant/runs)))
```

 Key point

If you can find a way to make R carry out the appropriate analysis for your question, you can use it in a simulation to estimate power.

Summary points

- Increasing the number of research questions in your study will complicate your power analyses; but all the principles introduced for simpler studies still apply.
- You can evaluate the powers of different designs by simulation for each of several hypotheses separately, with trivial changes to your computer code.
- No design is likely to offer exactly the same power for all these hypotheses.
- However, ranking these hypotheses in terms of their importance to your study can help you find a design that offers sufficient power for all the hypotheses of prime importance.
- Having settled on such a design, you can evaluate whether the power available for testing ancillary hypotheses is sufficient to continue to include them in your study.

Further reading

All of our favourite statistical textbooks give good coverage of both types of analysis discussed here, but we will also shamelessly point you to a paper by one of the authors that discusses approaches to comparing between levels of a factor:

Ruxton, G.D. and Beauchamp, G., 2008. Time for some a priori thinking about post hoc testing. *Behavioral Ecology*, 19(3), pp.690–693.

A couple of other papers we like on this issue are:

Glickman, M.E., Rao, S.R. and Schultz, M.R., 2014. False discovery rate control is a recommended alternative to Bonferroni-type adjustments in health studies. *Journal of Clinical Epidemiology*, 67(8), pp.850–857.

Nakagawa, S., 2004. A farewell to Bonferroni: the problems of low statistical power and publication bias. *Behavioral Ecology*, 15(6), pp.1044–1045.

Discussion question

7.1   In our pesticide example, if our primary interest is in whether the novel procedure offers an improvement over the traditional approach, do we really need the no-pesticide control group?

8 APPLYING OUR SIMULATION APPROACH BEYOND NULL HYPOTHESIS TESTING: PARAMETER ESTIMATION, BAYESIAN, AND MODEL-SELECTION CONTEXTS

Learning objectives

By the end of the chapter, you should be able to perform a simulation analysis in the contexts of:

- Precision of a parameter estimate.
- Bayesian statistics.
- Model selection.

The way we do statistics is changing. Back when we were undergraduates, null hypothesis testing was the absolutely dominant form of statistical inquiry. It still remains important to a lot of researchers in many different areas of the life sciences, and we think this will remain true for at least the next decade. That is why we have couched most of this book in the context of null hypothesis testing. But researchers now have a wider toolkit of statistical approaches in common use alongside (and in some cases in preference to) null hypothesis testing. Happily, everything we have discussed thus far about a simulation approach to power in the context of null hypothesis testing transfers very naturally to these other situations, as we will demonstrate in this chapter. The other contexts we have in mind are where you want to estimate the size of a parameter or select from a number of alternative models.

8.1   Likelihood of obtaining a specified precision of parameter estimation

Sometimes the purpose of your study is to obtain a reasonable precise estimate of something. Imagine you are a tomato farmer and your friends report good results from a particular plant food that you haven't used before. It makes sense to run an experiment to explore how much more yield you get from different application rates of this food. You need a sufficiently precise estimate of this in order to evaluate whether this food seems as though it might offer you an economic advantage; so you might want to ensure that your experiment has a good chance of delivering a specific precise estimate of the increase in yield. As another example, it is already well established that regular exercise can offer a broad range of health benefits. If one of us organized an experiment involving incentivizing some staff at our university to take daily exercise for 12 months, then we could expect that this would probably show that exercise led to a reduction in time missed from work through ill-health. The real value of the experiment would be to obtain a reliable estimate of the strength of this effect. Such a reliable estimate would be needed for the university management to help them decide whether they saw economic value to them in supporting such incentivization continuing beyond the scope of the experiment. Thus, we should be interested in ensuring that our experiment delivers a reasonably precise estimate.

Let's use a worked example to explore simulation analysis in the context of precision of parameter estimation. A well-known phenomenon in animal physiology is that resting metabolic rate across different broadly similar species of animals varies as mass to an exponent close to 0.75. That is, if we had two species of rodent, one of which had a mass ten times that of the other, we would not expect the larger species to use up energy to fuel its routine metabolism at ten times the rate of the smaller one, but rather only as five or six times the rate ($10^{0.75} = 5.62$). Although the value of this mass exponent seems close to 0.75 in studies of mammals, it tends to be a little lower (around 0.72) in studies of birds. Imagine that this set us thinking that it might be rather interesting to explore this relationship in bats—to see whether they conform more to the general mammalian pattern or whether selection for flight causes them to be more similar to birds.

The experiment we would perform to explore this is conceptually simple. We select several species of bat that cover a broad range of individual adult body masses. Kitti's hog-nosed bat (*Craseonycteris thonglongyai*) is the smallest bat, weighing only 2 g. It's not quite clear which of a number of bat species has the largest individuals, but it might well be the great flying fox (*Pteropus neohibernicus*) with masses up to 1450 g. There are 1200 species of bat to choose from. For each of our study species, we sample a number of adult individuals and for each individual we record their mass. Next, we place them separately in a chamber where we can monitor oxygen uptake and convert their lowest level of uptake to a basal level of metabolic rate. We can then take averages across individuals in a species and plot basal metabolic rate against mass with a point for each species, fit a line of the appropriate form, and evaluate the parameter that describes variation with mass in that model (we will call this parameter 'the exponent'). Now you could think of this study in terms of null hypothesis testing, where we are testing the null hypothesis that this parameter value is 0.75. However, we are not just interested in whether or not this parameter is or is not 0.75. We are interested in its actual value, and in comparing that value with both 0.75 and 0.72. So it is natural to think of our study as primarily aiming to estimate this parameter value with what we deem to be an acceptable level of precision. Our simulation approach can still be useful here.

The important issues when estimating a parameter are that you want your estimate to be unbiased and precise (we defined and discussed bias and imprecision in Chapter 3). Avoiding bias is primarily about good study design and careful statistics, improving precision is about the 'study effectiveness' issue that is our main interest here. The more effective your study, the more precise your parameter estimate will be. When you actually carry out statistical analysis of the bat study described above, you will probably get an estimate of the parameter of interest (the exponent of the relationship between BMR and mass) and a standard error associated with that estimate. The smaller the standard error, the greater the precision. So it might be rational to set some bound on the precision you feel is needed in your estimate. You could then use a very similar simulation approach to the one we took for power analysis to decide on the size of experiment needed to give you confidence that the estimate your experiment gives will probably be sufficiently precise.

In our case, we want a value that is precise enough that we can say whether it seems more like 0.75 or more like 0.72. Very crudely speaking you can think of the standard error as an estimate of the standard deviation of the distribution of predicted values you would get from lots of replicate samples. And if these estimates were normally distributed then we might expect about 95 per cent of them to fall within a range of four standard errors (since for a normal distribution 95 per cent of values lie within two standard deviations either side of the mean value). Thus if we had a target standard error of, for example, 0.005 then the 95 per cent confidence interval that we might construct to illustrate our uncertainty in our estimate of the exponent would be four standard errors (i.e. 0.02) wide, and so not be so wide as to include both the values 0.72 and 0.75 (see Fig. 8.1(a)).

In this situation, we would be able to say something useful about whether our data suggest that bats are more like other mammals or more like birds with regard to how individual basal metabolism changes with mass. However, if our estimate was less precise (such that both 0.72 and 0.75 seemed plausible values for this parameter on the basis of our data) then our conclusions would be much less interesting (see Fig. 8.1(b)). Thus we might decide that the target of our simulated experiments is to help us design an experiment that will give us an 80 per cent chance of producing a standard error for our estimate that is less than 0.005 (we might call this target the characteristic precision). In the next section we will look at the mechanics of exactly how to do that by our now-familiar simulation method.

Fig. 8.1 (a) In this scenario, where the target standard error is 0.005, the 95% confidence interval (the black line with arrows at each end) that we might construct would be four standard errors (i.e. 0.02) wide. Here, with a mean of 0.745, only one of our hypothesized values of 0.72 and 0.75 (shown by dotted lines) is included in the interval. With such a small standard error, we can be assured that the 95% confidence interval will not include both 0.72 and 0.75, regardless of the mean value. (b) In this scenario, where the target standard error is 0.01 but our mean is still 0.745, our 95% confidence interval is now 0.04 wide (0.01 multiplied by four). This less precise estimate means that both our hypothesized values of 0.72 and 0.75 seem plausible for this parameter on the basis of our data.

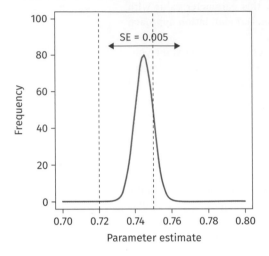
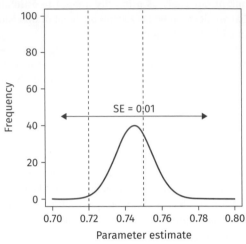

> ### 💡 Key point
>
> It can be useful to transfer our understanding of how to estimate statistical power to the goal of obtaining a specified precision for a parameter estimate.

8.1.1 Simulating our bat metabolism experiment

Conceptually, to estimate the likelihood of achieving at least some specified precision of a parameter estimate, we approach things in the same way that we have done throughout this book, by simulating many repeats of proposed experiments and studying the diversity of possible outcomes. However, whilst in the book so far we have focused on the p-values generated by our imaginary experiments, here we will change our focus to other statistical output. Let's take that approach with our bat experiment.

 To generate simulated data, we first need to select a sample of masses (in grammes) for our bat species. We decide to select these from a log-normal distribution with mean 50 and standard deviation 80. See the Appendix for details of this distribution, but briefly this is a handy distribution when you want to select real numbers (rather than integers) that can never be negative and which tend to have a skewed distribution such that most values are small but occasional large values are possible. A sample of 100 values is shown in Fig. 8.2.

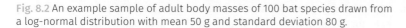

Fig. 8.2 An example sample of adult body masses of 100 bat species drawn from a log-normal distribution with mean 50 g and standard deviation 80 g.

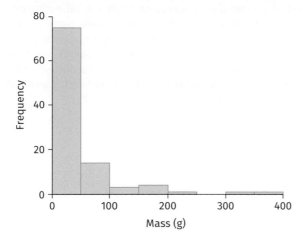

There is an issue with biological realism here, because the log-normal distribution can generate really tiny values that are unrealistically small for a bat. As a crude way to avoid this, we add 3 grammes to all the values we generate from this distribution.

Now for each of the mass estimates we need to come up with a corresponding estimate of resting metabolic rate (R). From the literature, we find that a large study of mammal species came up with the following relationship, between mass M in kg and R (in watts):

$$R = 3.34M^{0.75}$$

so we could use that to calculate our estimated values of R (after converting our sampled masses to kg). However, we would not expect all the values to lie perfectly along the line, so (based on previous studies) we decide to modify each R value by a percentage drawn from a normal distribution with mean zero and standard deviation 0.1. When we have a sample of paired mass and metabolic rate measurements (M and R) we want to use them to estimate the value of the exponent parameter and its standard error—so we want to fit a model of the form:

$$R = aM^b$$

Where a and b are constants, and we focus on the estimation of b. One common way to fit these models is to take logarithms of both sides of the equation, then from simple rules of logarithms we get:

$$\log(R) = \log(a) + b\big(\log(M)\big).$$

This means that if we plot the logs of our masses on the x-axis and the logs of our metabolic rates on our y-axis then we expect the relationship to be a straight line with gradient b, so we have reduced our model fitting task to fitting a straight line and estimating the gradient. Thus, our task actually boils down to one that is very similar to the kinds of simulation we were running in Chapter 6. The main difference is that once we use the R lm function to fit our statistical model, we don't pull out the p-value, but instead we pull out our estimate for the standard error of our gradient parameter. When we did that with one set of simulated data involving 50 individual bat species using the code below, we estimated b to be 0.766 with an associated standard error of 0.028. This standard error is quite a bit bigger than we would like (remember we decided to aim for 0.005). So we increased the sample size to 100 species. (Before moving on, there are a few handy R functions introduced in the code for this chapter, take the time—perhaps with a bit of internet searching—to make sure you are happy with all the code for this chapter.)

```
m <- 50
s <- 80
location <- log(m^2 / sqrt(s^2 + m^2))
shape <- sqrt(log(1+(s^2 / m^2)))
masses <- 3+(rlnorm(n = 50, location, shape))
M <- masses/1000
R <- 3.34*(M^0.75)
E <- rnorm(50, 0, 0.1)
R <- R*(1+E)
logM <- log(M)
logR <- log(R)
summary(lm(formula = logR ~ logM))
```

Changing the sample size to 100 gave us an estimate of 0.754 and standard error of 0.010. This is still larger than we would like so we tried a sample of 500 which produced an estimate of 0.749 and a standard error of 0.004, which is small enough. But we would like to know how lucky we were to get this value, so we repeated this for 10,000 samples to see what fraction of them gave standard errors less than or equal to 0.005. Using the code below we estimated this as 0.933.

```
m <- 50
s <- 80
reps <- 10000
counter <- 0
location <- log(m^2 / sqrt(s^2 + m^2))
shape <- sqrt(log(1+(s^2 / m^2)))
for (i in 1:reps){
masses <- 3+(rlnorm(n = 500, location, shape))
M <- masses/1000
R <- 3.34*(M^0.75)
E <- rnorm(500, 0, 0.1)
R <- R*(1+E)
logM <- log(M)
logR <- log(R)
results <- summary(lm(formula = logR ~ logM))
s_e <- results$coefficients[[4]]
if (s_e <= 0.005){counter <- counter+1}}
counter/reps
```

This seems higher than the 80 per cent we had decided upon, so it looks like we could reduce the sample size a little and still have a good chance of getting the precision we were looking for. We can explore how the probability of getting a standard error less than 0.005 varies with sample size. The results of such

Fig. 8.3 An investigation into how the effectiveness of the simulated bat experiment varies with sample size finds that as sample size increases between 400 and 500, so too does effectiveness. To achieve an 80% chance of producing a standard error for an estimate that is less than 0.005, we might need a sample size of around 480 species (red dashed lines).

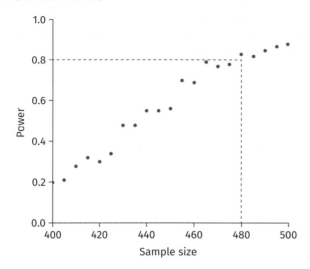

an investigation are shown in Fig. 8.3. The challenging news for us is that we might need a sample size of 480 bat species.

This seems a tall order (involving over one-third of all bat species). So we might explore whether there are ways we could change the experiment that might allow us to reduce sample size without compromising the precision of our estimate. Essentially, we might want to explore how we can reduce inherent variability. This might involve measuring more individuals per species, or standardizing on only using individuals of one sex, or perhaps changing how we measure metabolic rate to really ensure that we obtain the resting level and not a higher level due to the bats feeling stressed by the experimental procedure.

We could explore these changes by simulation, but our purpose here was to demonstrate the concept that we can use simulating experiments effectively in contexts other than testing a null hypothesis. Here we have shown how we might go about ensuring that when estimating a parameter value we have sufficient precision to allow us to draw useful biological conclusions.

But before we finish, we should make a number of confessions. By using the lm function to find our regression line, we are implicitly assuming that each point (each species) is statistically independent. This is a very dubious assumption, since some bat species in our sample could be closely related. There are much more complex statistical approaches that can take the relatedness between species into account. Also, the way we generated our data means that the variance in R increases with the mean, there are yet other statistical methods that cope with this better than the simple methodology we used. We skirted these statistical issues so we could concentrate on concepts related to power and precision. These concepts would carry over exactly if you adopted many statistical methods to give you a parameter estimate as long as that method also offers you some measure of the precision of that estimate (like a standard error or confidence interval).

Also, to avoid overwhelming you with details of bat biology, we have taken an approach to our notional experiment that probably few, if any, bat researchers would actually take. In an experiment like this, researchers would probably know the species they were going to use beforehand, and so actually use the mean masses of those species as fixed values in each simulation. This approach allows adding phylogenetic information to take relatedness into account, because you can only do this if you are using actual real bat species with an actual phylogeny. Even if the real experiment was going to pick randomly from all bat species, the researcher could number the bat species and pick random species, so that the masses represented real species with known phylogenies. These approaches could be simulated too, but would be a little more involved.

 Key point

It is conceptually and practically simple to use simulated experiments to estimate likelihood of obtaining a specified precision associated with a parameter estimate.

8.1.2 Selecting the criterion to explore in simulated experiments.

When testing a null hypothesis, power is uniquely defined as the probability of rejecting the null hypothesis when that null hypothesis is actually false. In other contexts, you have more flexibility in how you define the criterion you use to estimate the likely effectiveness of a study of particular design. In our bat-metabolism example, we selected the probability of having a standard error less than some limit that we set (0.005). We could just as easily have selected a different value for this critical standard error and/or some other measure of precision (for example, the 95 per cent confidence interval rather than the standard error).

This flexibility is a good thing; you can select the criterion which best fits with the biological question you are interested in. However, this does mean you have a little bit more work to do. Firstly, you need to think about what criterion would work best for you. Secondly, you have to make sure you state the criterion you are using clearly when you are presenting results of a set of simulations, and you have to justify this choice when presenting your results to others. We hope we have demonstrated in the bat-metabolism example that all this does not really require much extra work.

This flexibility does also allow you to explore the likely effectiveness of your experiment in a little more detail. Consider again our worked example, where we might have decided to use a sample size of 480 in the expectation that this gives us an 80 per cent chance of having the standard error within 0.005. This means we have a 20 per cent chance of the standard error being greater than that value. We can easily use our simulation approach to explore how much greater the standard error might be. We simulated 10,000 bat experiments with a sample size of 480 and in Fig. 8.4 we show the distribution of standard errors, the largest standard error was 0.00575, which would not be utterly useless for our purposes. So in this case, our additional exploration should even further

Fig. 8.4 The distribution of standard errors from 10,000 simulated bat experiments with a sample size of 480. Given that the largest standard error was 0.00575, we can feel confident that this sample size will produce useful data.

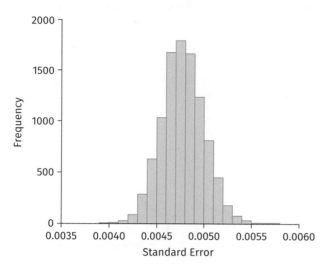

strengthen our confidence that our experiment should yield useful data. More generally, the flexibility associated with simulation analysis has the added value that we can usefully really explore the whole range of likely outcomes of our experiment to help us select the design that feels best to us.

 Key point

You have flexibility in how exactly you choose to define the criterion you use to estimate the likely effectiveness of a study of particular design; you can use this flexibility to your advantage.

8.2   Translating the concept of power across to Bayesian analysis

Another set of statistical tools being increasingly commonly used are those that adopt a Bayesian approach. If you are reading this section we can probably safely assume that you have a good understanding of Bayesian techniques. Essentially, these techniques are about exploring how the data you collect influences the distribution of values of a parameter of interest that you consider plausible.

Let's take a simple example. Suppose you are conducting an experiment to compare a single group of treated individuals to a control group. If you were null hypothesis testing you would probably carry out a t-test to see whether your data suggests the samples were drawn from populations with different mean values. As a Bayesian your parameter of interest would also be the difference between the means of the populations, but instead of a p-value, your analysis would deliver you a distribution of this difference based on your data (and any other prior assumptions that you chose to make). You would then use this so-called posterior distribution to draw conclusions about your experiment.

Fig. 8.5 Two-panel figure with posterior distributions that peak at the same point, but vary in their breadth.

Fig. 8.5 Two-panel figure with posterior distributions that peak at the same point, but vary in their breadth.

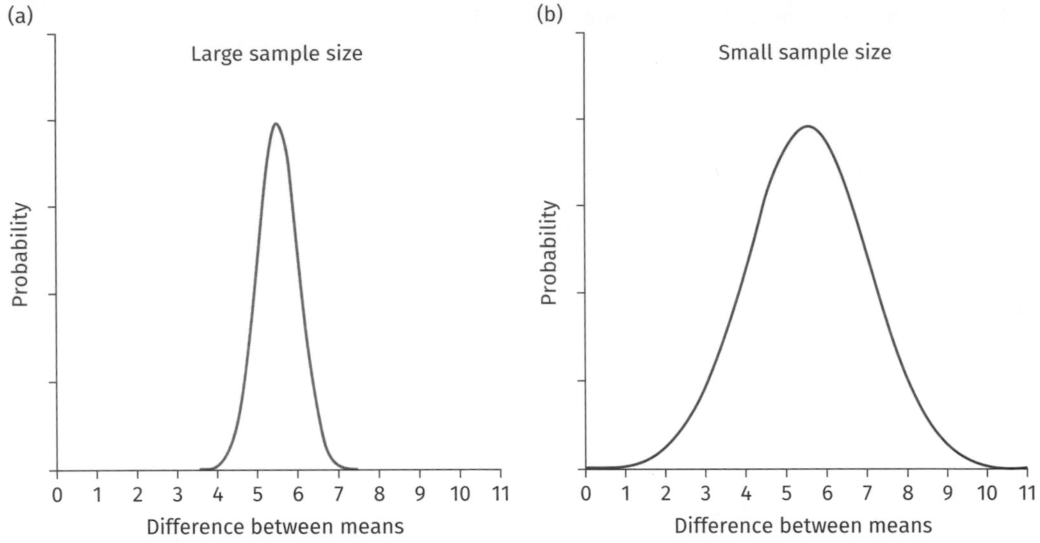

Figure 8.5 shows the posterior distributions of two runs of this study carried out by different researchers, each with a different sample size. Both distributions peak at very similar values, and so both studies suggest that the most likely difference is about five or six. However, the distribution around this peak is a measure of our uncertainty in our estimate of the difference. Experiment 1 (the left-hand panel) has produced a fairly tight distribution, whilst Experiment 2 (the right-hand panel) has produced a much broader distribution, and so greater uncertainty in the difference. Intuitively, you should be able to see that Experiment 1 is the better experiment. So to estimate the power of our experiment from a Bayesian perspective we can use the same approaches of simulating imaginary experiments that we have used so far in this book. As in the previous section, the information we would extract from each imaginary experiment is a measure of the uncertainty of our parameter estimates.

Earlier in this chapter we discussed the standard error as a measure of the precision with which we can estimate a parameter. A very common use of standard errors associated with estimated values of a parameter is to construct a confidence interval. An approximate rule of thumb for many situations is that the 95 per cent confidence interval spans the region from the estimated value minus twice the standard error to the estimated value plus twice the standard error; but there is a whole body of theory on how to estimate confidence intervals more precisely in different situations. However, the smaller the standard error the narrower the confidence interval will be. The important thing for us here is that the confidence interval is a measure of the range of possible values for the population parameter of interest that seem compatible with our sample data. The meaning of the 95 per cent confidence interval is that if we construct such confidence intervals based on repeated samples, then in 95 per cent of cases the confidence interval will encompass the correct population value. Very loosely speaking then, the narrower the confidence interval the more precise the estimate is considered to be.

The reason we are explaining standard errors and confidence intervals here is that Bayesian statistics have an analogous concept—the credible interval. This has a more straightforward interpretation than confidence intervals: there is a

95 per cent chance that the population value of the parameter of interest lies within the credible interval. Thus, the width of the credible interval can very much be interpreted as a measure of the precision with which your experiment is able to estimate the value of the parameter of interest. As such, exactly the same approach as that discussed with standard errors could be taken with credible intervals. Thus, our simulated-experiments analysis is entirely applicable to Bayesian statistics—with the aim being to design an experiment that allows the parameter of interest to be estimated with a particular precision. This precision can naturally be expressed in terms of the size of the credible interval.

There is also a Bayesian analogue of a p-value called the Bayes factor. The Bayes factor (generally denoted BF_{10}) is the likelihood of obtaining the observed data if the alternative hypothesis (H_1) were true divided by the likelihood of obtaining the observed data under the null hypothesis (H_0). The larger this factor the more the data offers relative support for H_1 over H_0. As a rule of thumb BF_{10} values above three are often described as offering substantial support for H_1 (relative to H_0); and values above ten, strong support. If you can find a statistical procedure that provides a value of BF_{10}, then we are sure you can see how you could apply our simulation approach to evaluate the probability of obtaining a BF_{10} value above ten (say) from a given design of experiment. (Just be on your toes—some packages will report BF_{01} values, but these are just the inverse of BF_{10}.)

There is not much more that needs to be said about Bayesian statistics in the context of simulated-experiments analysis. This is a short section not because we want to give the impression that such simulations are of little importance to a Bayesian statistician. Quite the reverse—we consider that it should be just as important to a Bayesian as to a frequentist statistician interested in null hypotheses. However, happily the concepts we have discussed in the context of power transfer very naturally from one statistical philosophy to the other. The way we would simulate our experiments is exactly the same with either approach; the only difference comes in the analyses you apply to our imaginary data, and the information to extract (e.g. the p-value or SE versus the Bayes factor or credible interval). Indeed, we can even imagine situations where, faced with alternative statistical approaches to analysing an experiment, you might use your imaginary worlds to work out which would give you the most information about the question you are trying to answer. We have chosen to write this book mostly in terms of null hypotheses and only briefly mention that the concepts transfer naturally to the Bayesian world—in a later edition we might decide to take entirely the opposite approach.

 Key point

A natural way to apply our simulated-experiments analysis to Bayesian statistics is to set a bound to the width of the credible interval associated with the parameter you are seeking to estimate, or a critical value for the Bayes Factor.

8.3   Using simulations to help with model-selection approaches

Model selection is another approach to statistics that lends itself naturally to simulation analysis. You might also see this technique described as an information-theoretic or likelihood-based approach. Essentially this involves fitting a number of models to your data, then using some measure of how well the different models fit the data to say something about the relative support that the data offers to the different models. Often the Akaike Information Criteria (AIC) is used to quantify how well each model fits the data, with the model producing the lowest AIC being considered the best model. A common rule of thumb is that models whose AIC is within 2 of that of the best model (i.e. the model from your set with the lowest AIC) are considered to be as well supported as the best model by the data; those that are within 2–9 are moderately supported in comparison, and those that differ by more than 9 from the best model can be considered as having very little support in comparison to the best model. You will find other measures used (such as the corrected AIC or Bayesian Information Criterion). You will also find a variety of different rules employed for using the measure of fit to describe the level of support that the data offers to different models. Again, analysis by simulation can be useful in this context too; let us demonstrate this with a conceptual example.

Imagine we are looking to design a long-term study of how fecundity changes with age in individually identifiable females of an albatross species. We suspect that fecundity increases with experience but shows no senescent drop-off in old age; however, we would like to know how many birds we should monitor and for how long in order to provide sufficient data to reliably confirm or refute our theory. Our plan would be that when we collect the data we fit it to four models that make the differing assumptions about fecundity: it is: (i) unaffected by age; (ii) increases initially with age then levels off; (iii) remains constant initially but declines in later years; or (iv) increases initially, then levels off in middle years before declining in old age. We intend to compare these four models and use their AIC values to decide which model fits the data best. However, we would like to design the study so that we have a high chance of obtaining an unambiguous answer as to which of these four possible models is the best description of reality. We might quantify this aim as demanding an 80 per cent chance that the AIC of the best fitting model is at least 2 less than any of the other three models. We can then carry out a simulation analysis under a range of assumptions for: (a) average fecundity; (b) strength of experience effect; (c) strength of senescence effect; and (d) mortality rate, in order to come up with an estimate of the appropriate sample size to use in our study to give us a good chance to identify which of our four candidate models is clearly the strongest supported. Such a simulation analysis seems particularly valuable for a study such as this that is likely to take a long time and be resource-intensive.

 Key point

Simulation analysis also applies naturally to model selection.

8.4   Exploiting your freedom in how you define study effectiveness

In the last section on model selection, just like for parameter estimation and Bayesian analysis, we can see that the simulation approach advocated in our book perfectly lends itself to evaluating the likely effectiveness of any study design just as naturally as with null hypothesis testing. The main issue in all these cases is that you have more flexibility in how you determine the criterion by which you quantify effectiveness. In our example above we could have defined this in lots of different ways other than the probability that the AIC of the best fitting model was at least 2 less than that of any of the other specified models. This flexibility should be seen as a good thing, not a problem. It requires you to think before you start the study about how you will analyse the data and importantly how you would interpret different potential outcomes of that analysis. The most relevant way to define effectiveness in your study should come very naturally from such deliberation. This is good news for you in a number of ways. It is good practice to think about your analysis and interpretation beforehand when you have more time and before you are actually committed to an expensive and time-consuming data-collection exercise. So deciding how you are going to define effectiveness doesn't really involve much extra work on your part and encourages you into the good practice of thinking about data analysis before embarking on collecting the data. The payoff for this is that you get to perform your simulation analysis based on a definition of effectiveness that seems exactly in tune with how you expect to interpret the results of your data analysis. This flexibility is also a reason why you are glad you adopted the simulation approach of this book and not some off-the-shelf package—any package is going to struggle to accommodate this flexibility.

 Key point

Thinking carefully about how to define the effectiveness of a study in these contexts has the ancillary benefit of encouraging you to think about how you will interpret the outcome of your statistical analysis at the outset of your study.

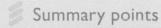 Summary points

- It can be useful to think about the effectiveness of a study in terms of the likelihood of obtaining a specified precision associated with the estimated value of the parameter that your data collection aims to estimate.
- In order to estimate the likelihood of obtaining a specified precision associated with a parameter estimate, it is conceptually and practically simple to use an analogous simulation approach to the one we used when power was used to define effectiveness. You have a choice in how you define effectiveness in this context and you can make this choice work to your advantage.
- A natural way to apply simulation analysis to Bayesian statistics is to set a bound to the width of the credible interval associated with the parameter you are seeking to estimate, or a bound on the Bayes factor.
- Simulation analysis can also easily be applied to the statistical technique of model selection.
- Thinking carefully about how to define effectiveness of a study in these contexts has the ancillary benefit of encouraging you to think about how you will interpret the outcome of your statistical analysis at the outset of your study.

Further reading

Burnham, K.P. and Anderson, D.R., 2004. Multimodel inference: understanding AIC and BIC in model selection. *Sociological Methods & Research*, 33(2), pp.261–304. [A very highly cited introduction to model selection.]

Cumming, G., 2013. *Understanding the New Statistics: Effect Sizes, Confidence Intervals, and Meta-Analysis*. Routledge. [A non-intimidating introduction to parameter estimation with lots of practical advice.]

Gardner, M.J. and Altman, D.G., 1986. Confidence intervals rather than P values: estimation rather than hypothesis testing. *British Medical Journal (Clinical Research Edition)*, 292(6522), pp.746–750. [An easy introduction to the attractions of parameter estimation as an alternative or complement to null hypothesis testing.]

Jarosz, A.F. and Wiley, J., 2014. What are the odds? A practical guide to computing and reporting Bayes factors. *The Journal of Problem Solving*, 7(1), p.2. [A very readable introduction to the Bayes factor.]

Kruschke, J.K. and Liddell, T.M., 2017. Bayesian data analysis for newcomers. *Psychonomic Bulletin & Review*, 25(1), pp.155–178. [As gentle an introduction to Bayesian methods as you will find.]

Manly, B.F., 2006. *Randomization, Bootstrap and Monte Carlo Methods in Biology*. Chapman and Hall/CRC. [This is one of our favourite books on using simulation methods generally, but it is very clear about parameter estimation and its value.]

Symonds, M.R. and Moussalli, A., 2011. A brief guide to model selection, multimodel inference and model averaging in behavioural ecology using Akaike's information criterion. *Behavioral Ecology and Sociobiology*, 65(1), pp.13–21. [Another popular and really clearly written guide to model selection.]

Discussion questions

8.1   Why would we not expect all the bat species to fit perfectly on the same relationship between mass and resting metabolic rate?

8.2   How might we improve the effectiveness of the bat experiment by other means that might allow us to reduce the sample size?

8.3   How might you incentivize university employees to take regular exercise?

8.4   Why do we measure more than one individual per bat species?

APPENDIX: SOME HANDY HINTS ON SIMULATING DATA IN R

This guide is very far short of exhaustive, but we hope covers quite a few things you will find handy. Some of these issues have been covered in our main chapter, but we thought it would be good to cover them again here, so you can find them easily. If there are issues you can't find here, then remember that one of the attractions of R is that it is so widely used that searching online for anything you are unsure about should quickly get you to a solution.

Clearing out R

Whilst technically not directly involved in simulating data, we are going to start off with an extremely useful line of code:

```
rm(list=ls())
```

If you run this line of code it clears out R's memory of all that has gone before. So any variables you have assigned or lists and data frames you have you have created will be removed. If you are working in RStudio, and you run this line of code, you will see that all the things in your Environment window will disappear. This is actually the opposite of simulating data, so why would you want to do it? Well, when you start working on a new script, you want to make sure that there is nothing hiding away in R's memory from any previous script you might have been working with. If you include this line as the first line of code in your script, you will ensure that you are working with a blank slate. We have added this line to every script in this book.

Naming

We can assign a name to a number or a list of numbers in R (we use the word 'list' but you will also—probability more properly—encounter the word 'vector' used in this context). We do that using code like that shown below. You could think of the construction <- as an arrow pointing from the thing we want to identify with its name. R likes you to describe a list of numbers by separating them with commas, enclosing them in brackets and placing a c at the front—you can think of the c as standing for connect or combine—as it connects or combines all these numbers into a list.

```
Nick <- 25
Graeme <- c(34,4,67)
```

Remember that R is case-sensitive, so it would assume that **Nick** and **nick** are different.

Names can be any combination of letters, numbers, underscore, and full stops, but you should start with a letter. So all these are fine:

```
Nick
Nick_Colegrave
Nick25
Nick.Colegrave
Nick.Colegrave_25
```

Names should be long enough to be intelligible and no longer. Remember you might end up having to look back at your code in six months' time, in which case **Treatment4Yield_July** is much easier to understand than **T4YJ**.

Drawing samples

As you know by now, one of the key ingredients in simulating imaginary experiments is drawing samples from populations with characteristics that you have defined based on your biological knowledge. Luckily, R makes this easy for us since it has a whole range of functions that can draw samples from a variety of different kinds of population.

Drawing samples of categorical data

Sometimes the variable we are interested in generating data on can exist in only one of a few categories: examples include sex, infection status, and blood group. We can sample from such distributions easily in R using the `sample` function.

For example, imagine that the sex ratio of the population you want to sample from is 50:50 and you want to draw a sample of ten individuals selecting the sex of each. The code below does just that:

```
dataset <- sample(c(0,1), size = 10, replace = TRUE)
```

We have coded the sexes as 0 and 1 for males and females respectively. And we tell the function `sample` that it should draw from the list of choices 0 and 1 for each of the ten individuals chosen. (Remember that R likes a list to be separated by commas, enclosed in parentheses, and preceded by a c.) Unless you tell R otherwise it will assume that all of the possibilities can be selected from the list with equal probability. We have to stipulate that we are sampling with replacement, so that the sex of a sampled individual is not influenced by the position it takes in our sample (remember also that R cares about case, so it is **TRUE** not **True**).

We do not have to have only two options, if you wanted to sample throwing a six-sided die seven times then you could use the code below:

```
dataset <- sample(c(1:6), size = 7, replace = TRUE)
```

Notice here that we have used another R shortcut: 1:6 means the same as 1,2,3,4,5,6 but takes less time to type, and offers fewer opportunities for typos! But what if the probabilities of the different levels of the category are not equal. Imagine that individuals in the population can either have a disease (coded as 1) or not have the disease (coded as 0). Further imagine that 23 per cent of the population has the disease, and we want to draw a sample of 12 individuals and explore disease status of individuals in that sample. Code we could use is:

```
dataset <- sample(c(0,1),size = 12, replace = TRUE, prob =
c(0.77,0.23))
```

Here for each item in our list we specify a corresponding probability of being selected, so the probability of 0 being selected is 0.77, and of 1 is 0.23. The probabilities should of course add up to 1, and there should be a probability associated with each item on the list of possible selections (two in our case).

Permutations and indexing

Quite often with categorical values, we do not want a sample but a permutation. The best way to explain this is by example. Imagine that we have 40 plants labelled 1, 2, . . . , 40; and we want to randomly assign these into four groups of ten, plants in each group will be given a different herbicide treatment (labelled A,B,C,D). We can simulate this easily, because sample(n) generates a random permutation of the numbers 1 to n. so the code below achieves what we want:

```
perm <- sample(40)
treatmentA <- perm[1:10]
treatmentB <- perm[11:20]
treatmentC <- perm[21:30]
treatmentD <- perm[31:40]
```

This introduces square brackets as how R indexes specific items on a list. The name treatmentA is a list of ten unique numbers between 1 and 40. If we wanted to know all these numbers we could just type **treatmentA**, but if for some reason we wanted to know the seventh number on that list, we could get that by typing **treatmentA[7]** into R.

Such individual indexing can be handy, as we will show in the example below. Imagine we want to sample 25 individuals from a population with respect to their IQs. Imagine also that we know that sex influences IQ in our population. (This is a hypothetical example, we are absolutely not claiming that in actuality we have any knowledge of a link between sex and IQ.) Specifically, the female IQs are normally distributed with a mean of 120 and standard deviation of 15 and males are normally distributed with a mean of 115 and standard deviation of 20. We also know the population sex ratio is skewed to 60 per cent females. We could get our sample using the code below:

```
#set the sample size at 25.
sample_size <- 25
#set up a list called IQ that is currently all zeros, but
will be
# populated with values later.
IQ <- rep(0,sample_size)
sex_sample <- sample(c(0,1), size = sample_size, replace =
TRUE, prob = c(0.6,0.4))
for (j in 1:sample_size){
if (sex_sample[j] == 0){IQ[j] <- rnorm(1,115,20)}
else{IQ[j] <- rnorm(1,120,15)}
}
```

There are lots of handy thing to pick out of this. So let's take it slowly.

We first of all decide to call a variable **sample_size** and assign it the value 25. This is good practice, because you will see that we use this at numerous points throughout the code. If we wanted to change the sample size in our simulation then using our technique of giving the value a name at the start of the code and using that name throughout the code, makes things easier. To change the value of that parameter we need only change it once in an easy-to-find place, rather than going looking for all the instances we need to change throughout the code.

It's good to write comments in your code to help you understand it. This is especially helpful if you return to a code after weeks, months or years. If R sees the symbol **#** at the start of a line then it just ignores that line—so that is how we mark comments.

The next thing we did was decide that we are going to store the final IQ values in a list called IQ, this list is going to be 25 entries long. Before we start we fill this list with 25 zeros using the **rep** function in R. Ultimately, we will write over all these zeros, so this is not actually required but is good practice for a number of reasons. It warns R to make space for this list in its memory before you start to fill the list and this makes the code run faster. It also helps keep your plans straight in your own head if you declare right at the start of your programme any lists of variables that you are going to fill using your code.

The next thing we do is generate the sexes of the 25 individuals in our sample. We need their sexes in order to generate their IQs. We do that inside a **for** loop, because we are going to have to look through all the 25 individuals in the sample, taking them one at a time.

The first time the **for** loop runs, j is set equal to 1, so sex_sample[j] will give us the sex of the first person in our sample. We have now used an extension to the **if** statement introduced earlier in the book to include an **else** statement, you can see that what we do is ask if the sex of the 1st individual is male, and if so we draw the IQ of the first individual in our sample from the male population distribution; else (logically, if that individual is female) we draw from the female population distribution. We then go around that loop again, but this time j will be 2 so we will be dealing with the second individual in our sample. We just continue looping round until we have filled the whole sample.

Finally, notice if you ask in R if two things are equal (as we do in our **if** statement), R likes you to have two equality signs (OK, we say likes, but we mean R requires you to have two equality signs!).

There are lots of alternative ways we could have solved this coding challenge. We have picked the way that we think is easiest to understand. There are quicker-to-execute and/or quicker-to-type alternatives, but you can explore these are your programming confidence grows. There are many tips to help you do this in the ESM associated with Chapter 4 of this book, and we encourage you to play around with them.

Generating counts

Imagine that you receive on average four items of mail each day, you would not expect to obtain exactly four each and every day, rather the number will fluctuate but will always be a non-negative integer (you will never get -3 items of mail, or 0.2). One way to generate simulated data on the numbers of items of mail you obtain every day is to make the following two assumptions: (1) the average rate is constant over time; (2) individual items arrive independently of each other. These seem reasonable assumptions for our mail example (except possibly around your birthday or Christmas). If we can make these two assumptions then the arrival of mail can be modelled as a Poisson variable. To obtain the simulated arrival of mail over 14 days we just need to specify the mean rate of arrival λ (in our case 4). Then we obtain our simulated dataset simply from:

```
mail <- rpois(14,4)
```

To give a more biological example, we might be able to specify the probability of a mutation occurring per base pair, and simulate the number of mutations in strands of DNA of specified lengths.

Generally, the restrictive assumption for applying a Poisson distribution is that events (mail arrivals or mutations) occur independently of each other. If you find you can't live with that assumption in the system you are simulating then the negative binomial distribution might well be the answer. The Poisson distribution has the property that the variance of generated values is equal to the mean, for cases where events are clustered together then the variance will be larger than the mean, and that is the situation that the negative binomial is ideal for handling. We generate n values from such a distribution in R using the following function:

```
Clustered <- rnbinom(n, size = r, prob = p)
```

Here, each of the n values is the number of failures generated in Bernoulli trials with probability of success p before a total of r successes are achieved. This may strike you as an odd formulation, but the key thing is that the mean of a long string of generated values will be $pr/(1-p)$ and the variance will be $pr/(1-p)^2$. So you can use the two parameters p and r to get the shape you want, given that the mean divided by the variance is simply $(1-p)$. There is an alternative way to specify the parameters of this function that you might find more intuitive:

```
Clustered <- rnbinom(n, mu = m, size = s)
```

In this case the mean of a large string of generated values will be simply m and the variance will be $m + m^2/s$. So now you can specify the mean value directly, then use s to change the variance—variance will increase with mean but will also increase for smaller values of s.

Generating samples of continuous variables

A uniform distribution

The simplest continuous probability distribution is a uniform one where all values are equally likely. The code:

```
Sample <- runif(100)
```

gives you 100 continuous values each independently drawn from a uniform distribution between 0 and 1. If instead you want the numbers drawn between 0 and some specified maximum, you just multiply your random numbers by the maximum of your desired range. So imagine you wanted to pick ten compass directions (in degrees) at random with all directions being equally likely—you could get these simply from:

```
Angles <- 360*runif(10)
```

The normal distribution

Another really common continuous distribution is the normal distribution—here data points tend to be clustered around the mean with greater and greater deviations from the mean being less and less probable (but still possible). We made use of this distribution extensively in the main text. To describe a normal distribution we need to specify the mean value and the standard deviation—which is a measure of how spread the data points tend to be around the mean—about 95 per cent of the data points will lie within two standard deviations of the mean. We might consider the heights of adult male humans in our population to be well described by a normal distribution with mean 70 inches and standard deviation 2.5 inches, so that on average men are 5 foot 10, but only the unusual one man in 20 is outside the range 5 foot 5 to 6 foot 3. We could sample 30 men from this population with the code below:

```
Heights <- rnorm(30, mean = 70, sd = 2.5)
```

The log-normal distribution

There are two issues that might turn you away from the normal distribution: (i) it is symmetrical about the mean; and (ii) it can potentially generate negative values. In our heights example there is a trivially small risk of heights that are too small to be plausible being generated. This risk is trivially low because of the relatively small standard deviation. However, if you want to be certain that only non-negative values are possible (even if some values are pretty small) and/or you expect the distribution around the mean to be asymmetric then the log-normal distribution might be a good choice for you. As an example, the distribution of comments posted on internet discussion forums are generally well described by a log-normal distribution. Of course, the number of comments cannot be negative. Most topics generate either no comments or small number of comments (so the mean might be quite low—say four or five), but a fraction of comments generate a very large number of comments (over 100), so the distribution is strongly asymmetrical. Notice that number of comments is clearly going to be a non-negative integer value, but the log-normal distribution will generate positive real number values—you can convert real numbers by rounding to the nearest integer using the round function in R. Like the negative binomial distribution discussed earlier, the parameterisation of the function rlnorm that we use to generate samples from a log normal distribution are a little non-intuitive unless you are a statistician. Basically, it uses two parameters that can be related to the mean and standard deviation of the distribution. But imagine we wanted to generate a sample of the numbers of comments generated by 12000 randomly selected topics, where we expect the underlying mean to be 4.8 and standard deviation to be 34.2, then the code below will do this for us:

```
m <- 4.8
s <- 34.2
location <- log(m^2/sqrt(s^2 + m^2))
shape <- sqrt(log(1+(s^2/m^2)))
comments <- round(rlnorm(n = 12000, location, shape))
```

Notice that we have introduced you here to a few more functions in R—sqrt() gives you a square root, log() gives you a natural logarithm, ^ gives you a power.

Generating correlated data

You might want to sample from two different variables with a specific correlation between them. For example, you might want to draw a sample from a range of people and measure two traits on each. For each individual you want data on their height and IQ. You assume that IQ is normally distributed with mean 120 and standard deviation 15, and height in cm is also normally distributed with a mean of 175 and standard deviation 5. However, you might also want to assume that there is a relationship between these two variables—specifically that taller people tend to have higher IQs—expressed as a correlation between the two variables of 0.6. (Again, we do not mean to imply that we actually believe taller people have higher IQ, it is a hypothetical example.) We can generate a sample of the heights and IQs of 10 such individuals using the code below.

```
library(MASS)
n <- 10
mu1 <- 220
mu2 <- 175
s1 <- 15
s2 <- 5
rho <- 0.6
mu <- c(mu1,mu2)
sigma <- matrix(c(s1^2, s1*s2*rho, s1*s2*rho, s2^2), 2)
mydata <- mvrnorm(n, mu = mu, Sigma = sigma)
myX <- mydata[,1]
myY <- mydata[,2]
myX
myY
```

There are a number of issues raised by this example. First of all, the core version of R is a tremendously flexible simulation, data manipulation, statistics, and graphing package—but it cannot do everything you might want. When you want to do something a bit specialist you often have to install a further package into R to increase its functionality. Here, when we web-searched generating data of this sort a lot of websites recommended the function mvrnorm that we use above, and indicated that this function was available through the add-on packaged MASS. Packages are free and generally easy to install from R, and you can find more details of how to install them using RStudio in the ESM associated with the introduction to this book. and there is always the web if you can't figure it out. You only have to install a new package once, R will remember it has this add-on even if you switch your computer off. However, we have to connect to any packages we have installed and that we want to use each and every time we open a new session in R. This is what we are doing with the command library(MASS).

The function produces what you can think of as a matrix of values with a row for every one of the ten individuals, column 1 is the IQs of the individuals, column 2 is their heights. In order to separate this into two lists of all the IQs and heights, we use the commands

```
myX <- mydata[,1] myY <- mydata[,2]
```

We freely admit we picked an easy case where both variables are normally distributed, you can simulate correlated values where one or more of the variables is non-normal—how exactly to do that for the situation you want will take a little web searching on your part.

GLOSSARY

accuracy—see **bias**

Akaike Information Criteria (AIC)—A measure used when comparing the support that collected data offers to theoretical models. AIC score rewards both model simplicity and how well the model fits the data.

background variation—see **inherent variation**

Bayesian—Bayesian inference is a statistical approach in which Bayes' theorem is used to update the probability for a hypothesis as more evidence or information becomes available (for example, from data collected in an experiment).

between-individual variation—see **inherent variation**

bias—a consistent tendency to either over- or underestimate (and see **imprecision**).

blocking factor—Some variable that can be measured on subjects, that is expected to influence **between-individual variation** and that can be used to group 'similar' subjects together as part of an experimental design, so that the influence of a factor of interest can more clearly be seen by making comparisons only within such groups (called blocks).

confidence interval—When estimating a parameter from a collected set of data, the confidence interval can be considered informally as a range of values for that parameter that appear to be compatible with the data. More formally, if confidence intervals are constructed correctly using a given confidence level (often 95 per cent) from an infinite number of independent data-collection exercises, the proportion of those intervals that contain the true value of the parameter will be equal to the confidence level.

covariate—A **factor** that can be measured on subjects and included in statistical analyses to explain a portion of **inherent variation**, so that the effect of the factor of primary interest can be seen more clearly

credible interval—In **Bayesian** statistics, an interval within which the value of the parameter we are interest in falls with a particular probability. Analogous to a **confidence interval** in frequentist statistics.

dependent variable—see **response variable**

descriptive statistics—see **inferential statistics**

effect size—We have defined a parameter as a property of the underlying population that we are interested in: the effect size is the quantification of that parameter.

experimental units—see **subjects**

external validity—People describe a study as having low external validity if it is based on a **sample** drawn from a narrower population than the one you are really interested in.

extraneous variation—see **inherent variation**

factor—see **independent variable**

frequentist statistics—An approach to statistical inference that underpins null hypothesis significance testing. Its key assumption is that any given experiment can be considered as one of an infinite sequence of possible repetitions of that same experiment, each capable of producing statistically independent results; with the results of the experiment that you actually performed being interpreted in that context.

imprecision—Imprecision adds errors to your measurements but in a way that is uncorrelated from one measurement to the next (and see **bias**).

inaccuracy—see **bias**

independent factor—see **independent variable**

independent variable—If we are interested in how one characteristic of our experimental subjects (variable A) is affected by two other characteristics (variables B and C), then A is often called the response variable or dependent variable and B and C are equivalently called independent variables, independent factors, or sometimes just **factors**.

individuals—see **subjects**

inferential statistics—Statistical treatment of data that involves making inferences about a wider **population** on the basis of a **sample** drawn from that population. In contrast, **descriptive statistics** describes and summarizes aspects of the sample.

inherent variation—Any variation in the response variable between subjects in our sample (between-individual variation) that cannot be attributed to the independent **factors** is equivalently called inherent variation, random variation, background variation, extraneous variation, within-treatment variation, or noise.

interaction—Predictors (**factors** or **covariates**) of a response variable are said to interact if the apparent effect of one is influenced by the value of the other. Human Height and Sex would be said to interact in predicting Weight if the slope of the relationship between Height and Weight differed between the Sexes.

Logistic Curve—An S-shaped curve, as the x-variable increases from a low value, y first increases slowly, but this response accelerates as x increases further becoming very steep, before saturating with ever-larger values of x.

main effects—Predictors (**factors** or **covariates**) of a response variable are said to show main effects if the apparent effect of one is not influenced by the value of the other. Human Height and Sex would be said to both show main effects in predicting Weight if the slope of the relationship between Height and Weight is the same for each Sex—but there are Sex-specific intercepts (c.f. **interaction**).

meta-analysis—Evaluation of an aggregation of previous studies all exploring the same scientific question to see whether patterns can been drawn from combining understanding from these studies.

model selection—A form of **inferential statistics** in which a number of pre-defined models of underlying processes are fitted to the data from an investigation and the relative successes of these models in capturing patterns in the sample data are evaluated.

noise—see **inherent variation**

normal distribution—A distribution of values commonly used in statistical theory—its properties include being symmetrical about its mean value, the probability associated with values peaking at the mean value, and that probability falling away with increasing distance from the mean following a bell-shape.

null hypothesis—For any biological question that we are interested in answering, the null hypothesis is usually the answer that nothing is happening.

null hypothesis statistical testing (NHST)—an approach to **inferential statistics** that focuses on generation of a **p-value**.

omnibus null hypothesis—When testing for the effect of a **factor** with more than two values, the omnibus null hypothesis is that all the values of that factor have the same effect.

p-value—The probability of obtaining a set of data at least as extreme as the data we have observed if the **null hypothesis** is actually true for the **population** that we are interested in.

pairwise comparison—When testing for the effect of the value of a **factor** with more than two levels, a pairwise comparison tests the **null hypothesis** that two specified value of the factor have the same effect. Such pairwise comparisons are often

implemented after the **omnibus null hypothesis** has been rejected.

parameter estimation—A parameter is simply any attribute of the underlying population that we are interested in. Parameter estimation uses information from the **sample** to estimate what range of values for that parameter, at a **population** level, might seem plausible in the light of its measured value.

participants—see **subjects**

population—When statisticians talk about a population they mean the collection of all the entities that could be evaluated to answer the question of interest (and see **sample**).

positive predictive value—The probability that, if a statistical test suggests that there is an effect, this effect actually exists (often denoted *PPV*).

posterior distribution—In **Bayesian** statistics, the distribution of probability across possible values of the parameter of interest is specified prior to the experiment, based on understanding of the system. Then data is collected and that data is allowed to influence the distribution, leading to the so-called posterior distribution.

power—see **statistical power**

precision—see **imprecision**

pre-study odds—(often denoted *R*) This is the odds before we collect any data in our focal experiment—hence 'pre-study'—that the **null hypothesis** in a given experiment is actually true.

proteus phenomenon—This is where a discovery of an apparently substantial and interesting effect is not backed up in subsequent studies that find less strong effects or no evidence of an effect.

proxy measurement—Measuring one variable as a proxy for another variable than the focal variable of interest, that is expected to be highly correlated with the proxy but which for some reason is more difficult to measure.

random variation—see **inherent variation**

repeatability—A measurement has high repeatability if you get the same score each time that you measure the same thing, and has low repeatability if your scores tend to be inconsistent.

replicates—see **subjects**

response variable—If we are interested in how one characteristic of our experimental subjects (variable *A*) is affected by two other characteristics (variables *B* and *C*), then *A* is often called the response variable or dependent variable.

sample—Scientists study an appropriate representative sample of a **population** of interest and then make use of statistics to try to make generalizations about the population that they represent (and see population).

standard deviation—a measure of the variability between items in a **sample.**

standard error—When a **sample** is used to make an estimate of a population-level parameter (like a mean value) the standard error gives a measure of how much deviation we might expect between the sample-based estimate and the value we would obtain from another equivalent sample. It is often used as an indication of the precision of our estimate.

statistical power—The statistical power of an experiment is one minus the likelihood of making a **Type II error** on the basis of the outcome of that experiment. To put it another way: the statistical power of an experiment is the probability that you will correctly reject the **null hypothesis** if in fact the null hypothesis is untrue for the **population** of interest.

subjects—Your scientific study will involve you taking measurements on a **sample** of 'things': in this book, where we need a name for the 'things' in a sample, we will generally call them subjects, but you might also encounter them called experimental units, individuals, replicates, or participants.

subsampling—In subsampling, we measure a number of randomly selected parts of each individual that we want to sample and take the mean of these as a representation of that experimental unit.

surrogate variable—see **proxy measurement**

test statistic—Statistical tests often produce some single summary of essential patterns in the data—this is called the test statistic.

Type I error—in **null hypothesis statistical testing (NHST)**, rejecting the **null hypothesis** when it is actually true.

Type I error rate—In **frequentist statistics**, this is the fraction of replicate experiments that would lead to the conclusion that the **null hypothesis** should be rejected when in fact the null hypothesis is true. In NHST this rate is equivalent to the significance threshold for the **p-value** set by the researcher.

Type II error—in NHST, failing to reject the **null hypothesis** when it is actually false.

Type II error rate—in **frequentist statistics**, this is the fraction of replicate experiments that would lead to the conclusion that the **null hypothesis** should not be rejected when in fact the null hypothesis is false.

winner's curse—If an investigation is low-powered, then because of this low power we will have a high probability that the experiment will fail to detect any real underlying effect. When it does reject the **null hypothesis**, this will be because the small real effect and the chance effects of sampling have worked in the same direction, producing what appears to be a larger effect. It is these apparently large effects that trigger the statistical test to return a low **p-value**. Thus, if we are only interested in effect sizes when the statistical test suggests that there is actually an effect, then the size of the effects that we look at will tend to be inflated. This phenomenon of inflation is often called the winner's curse in the statistical literature.

within-subjects design—a type of experimental design where each **subject** experiences different treatments sequentially.

within-treatment variation—see **inherent variation**

INDEX

Note to index: *f*, and *t* attached to page locators denote *figure*, and *table*